Current Topics in Microbiology and Immunology

Ergebnisse der Mikrobiologie und Immunitätsforschung

70

Edited by

W. Arber, Basle · W. Henle, Philadelphia · P. H. Hofschneider, Martinsried

J. H. Humphrey, London · N. K. Jerne, Basle · P. Koldovský, Philadelphia

H. Koprowski, Philadelphia · O. Maaløe, Copenhagen · R. Rott, Gießen · H. G. Schweiger,

Wilhelmshaven · M. Sela, Rehovot · L. Syruček, Prague · P. K. Vogt, Seattle

With 14 Figures

Springer-Verlag Berlin · Heidelberg · New York 1975

ISBN-13: 978-3-642-66103-7   e-ISBN-13: 978-3-642-66101-3
DOI: 10.1007/978-3-642-66101-3

Table of Contents

Theoretical Aspects of Structure and Assembly of Viral Envelopes

Herbert A. Blough and John M. Tiffany [1,2]

With 2 Figures

Table of Contents

I. Introduction

Viruses which can be seen in the electron microscope to have a coherent envelope structure are all found to contain lipid, and this has in the past

1 Division of Biochemical Virology and Membrane Research, Scheie Eye Institute, University of Pennsylvania School of Medicine, Philadelphia, Pa. 19104, USA. JMT present address: Nuffield Laboratory of Ophthalmology, Oxford University, Oxford, England.
2 Supported by grants from the National Cancer Institute (CA 15370); Damon Runyon Memorial Fund for Cancer Research (DRG-1254); National Eye Institute (EY 01067-02) and a contract from the U.S. Army Medical Research and Development Command (DADA 17-72C-2147).

been used as a means of classifying viruses, secondary to the type of nucleic acid they contain (COOPER, 1961). Another common feature which these viruses have more recently been found to possess is carbohydrate (other than that in their nucleic acid), and in all cases so far studied their distribution has been found to be asymmetric—the glycoproteins being found only in the exterior of the viral coat. The envelope may therefore be defined as a protective lipoprotein container for the genetic material, bearing, embedded in its surface, glycoproteins and glycolipids responsible for cell surface recognition of the virus and for the preliminary stages of infection (attachment and penetration). The virus is generally sensitive to agents such as detergents or lipid solvents, which disrupt the lipid region of the envelope. Virus assembly generally occurs at a specific cell membrane, where virally-specified materials are incorporated into the membrane to begin forming the envelope, followed by formation of a bud which eventually surrounds the viral core material.

The general field of virus assembly and structure is largely covered by a number of recent and more extensive reviews (LAVER, 1973; LENARD and COMPANS, 1974; SCHULZE, 1973) and we have recently published a review of many aspects of lipid involvement in viruses (BLOUGH and TIFFANY, 1973). It would be largely superfluous to attempt to cover the same ground in the present review; rather, we propose to consider methods of investigating structure and assembly, and to point out limitations which must be placed on their interpretation. We shall also consider the process of assembly, in particular those aspects yet to be elucidated by experiment, with the hope of suggesting aspects amenable to practical testing and perhaps of drawing attention to larger structural resemblances between different virus classes. We include some models derived from existing data on several viruses, together with estimates of the magnitude of forces likely to operate within the envelope to maintain a given structure.

The role of lipid in virus envelopes must be carefully considered, as well as the reliability of methods used to probe its functions, since it is intimately involved in the process of insertion of viral surface proteins into the template membrane, and bud formation. In recent years lipid has been variously considered to be passively incorporated, although with a structural function (e.g. KINGSBURY, 1972), selected from an available pool of preformed host lipids, or as a fortuitous space-filler. Since lipids are synthesized by the host cell, the effect of virus infection on host cell lipid metabolism is of obvious importance. We refer to a considerable amount of recent work from this laboratory which tends to disprove earlier ideas about the origins of viral lipids. Mention is also made in Section III of some aspects of biosynthesis of viral polypeptides and carbohydrates.

Many of the points raised or structures suggested have no experimental basis as yet, and indeed many workers in the field of virology are dubious of the value of model-building. However, considerations of the overall geometry of the envelope and its constituents have proved to be useful in determining the number of surface projections of the virion (TIFFANY and BLOUGH, 1970b),

Table 1. Enveloped viruses of vertebrates and some representative members[a]

Classes	Representative members
1. RNA viruses	
Orthomyxoviruses	Influenza (human, equine, avian, porcine) viruses
Paramyxoviruses	Newcastle disease, parainfluenza, mumps, measles [b] viruses
Alphaviruses (Arbovirus group A)	Sindbis, Semliki Forest, equine encephalitis, rubella [b] viruses
Flaviviruses (Arbovirus group B)	Yellow fever, Dengue, Japanese B encephalitis viruses
Rhabdoviruses	Vesicular stomatitis, rabies viruses
Oncornaviruses	Rous sarcoma, mouse leukemia (e.g. Rauscher, Moloney, Friend), mouse mammary tumor, visna viruses
Coronaviruses	Avian infectious bronchitis, mouse hepatitis, human respiratory viruses
Arenaviruses	Lymphocytic choriomeningitis, Lassa, Parana viruses
2. DNA viruses	
Herpesviruses	Herpes simplex, varicella, infectious bovine rhinotracheitis, Epstein-Barr, Marek's disease viruses
Iridoviruses	Amphibian polyhedral cytoplasmic (frog virus, FV3) virus
Poxviruses	Vaccinia, orf, smallpox, fowlpox, Yaba tumor viruses

[a] Modified from WILDY (1971).
[b] Tentative classification.

the number of different polypeptides in the virion and the volume they will occupy, and indeed in predicting the existence of an envelope structural polypeptide for influenza virus (TIFFANY and BLOUGH, 1970a). Model building has served a valuable function in determining the structure of many macromolecules, and of viruses possessing cubic symmetry, and if properly applied can offer considerable help in elucidating the structure of the virus envelope.

Table 1 is taken in part from BLOUGH and TIFFANY (1973) and indicates most of the major groups of lipid-containing mammalian viruses, as well as some others which have been extensively studied and which cast light on the general principles of virus structure. As with our review of viral lipids (BLOUGH and TIFFANY, 1973), attention centers chiefly on the myxoviruses, and many illustrative examples are drawn from the literature on this group. This reflects the intense study over a considerable number of years applied to every aspect of infection by these viruses, and the relative paucity of data for many of the other virus groups. In addition, we shall omit consideration of chemical or physical properties of the internal nucleoprotein of the virus, unless the conditions of viral assembly are such that it imposes a particular form on the envelope.

II. Methods of Investigation of the Viral Envelope

A. Information from Intact Virus Particles

The essential prerequisite for study of viral structure is purity of the virus preparation. The techniques used may depend on the system used for virus growth—i.e. whether the virus is harvested from tissue culture supernatant fluids, from allantoic fluid following growth in embryonated eggs, by disruption of host cells to release viral aggregates, etc. Unless host material can be rigorously excluded, the chemical composition of the virus cannot be accurately determined. Information derived from physical probes (e.g. electron spin resonance spectra of spin-labelled lipids incorporated into whole virus) may also be suspect if insufficiently-purified virus is used, since the structure of lipid regions of the virus envelope is possibly different from that in the host cell membrane fragments which are also present.

Primary information on assembly, release and structure of the intact virus particle is derived from electron microscopy. There are, however, many limitations to this technique and its derivatives, which should be borne in mind in interpreting the images obtained. The preparation of positively-stained thin sections involves fixation, dehydration, and embedding of the virus, and the use of solvents in these stages may remove lipid; this can be avoided by the use of water-miscible embedding media, but at the price of more frequent polymerization problems. The image also depends on the staining procedure used (COMPANS and DIMMOCK, 1969; NERMUT et al., 1972), and generally definition of ultrastructure in poor. The negative contrast technique has unique capability in resolving surface structure, but problems of interpretation may arise when superimposition of images from the upper and lower surfaces of the virus particle is seen (VERNON et al., 1972). Isolated subunits of the virus have been visualized by this method (LAVER and VALENTINE, 1969). Approximate molecular weights of subunits can be obtained from their dimensions, if a value is assumed for their partial specific volume. However this technique has not been extensively used in the determination of infrastructure (as opposed to surface morphology) because the envelope may disrupt or break up into small "rosettes" under the action of surface forces during drying, or following treatment with agents such as ether (HOYLE et al., 1961); there is no way of telling whether this disruption accompanies phase transitions of lipid components in the envelope. Other drawbacks of the negative-staining method include shrinkage and distortion of the virus (NERMUT and FRANK, 1971) and interaction between stain and specimen. Freeze-dried particles (with or without freeze-etching) require metal shadowing, which will add at least 15 Å to the apparent size of any structure such as surface projections and will increase the apparent smoothness of etched or fractured surfaces. It has the advantage that no fixation or solvent dehydration is required, and except for the thickness of the metal shadow, probably approaches closer to the original dimensions of the particle than any other method. This and related techniques are dealt with in more detail by ZINGS-

HEIM (1972). More complex techniques, such as identification of surface chemical groups by antibody or lectin binding, may also be combined with electron microscopy to show the site of binding (LAFFERTY, 1963; AOKI et al., 1970) and to indicate that the particles are still intact.

X-ray diffraction to determine the disposition of envelope lipids and proteins has been used successfully only on Sindbis and PM2 viruses (HARRISON et al., 1971a, b). Interpretation of the results for more complicated viruses will be extremely difficult. Even for these comparatively simple examples (which contain a very small number of envelope polypeptides), it is hard to distinguish between icosahedral and truly spherical structures. Fourier synthesis of diffraction data leads to a radial electron density distribution for the particle, from which the dimensions and thickness of the lipid region and surrounding protein-rich areas can be deduced. The question of whether protein protrudes through the lipid bilayer region in these viruses is not yet satisfactorily resolved, but X-ray diffraction techniques do not seem to give an unequivocal answer.

A variety of spectroscopic techniques can be used to investigate the state of mobility of the lipid regions of the intact particle. These involve adding a lipid-soluble substance to the virus, and comparing its characteristic spectrum with that of the same probe molecule in structurally-defined surroundings; e.g., one may use nitroxy-stearic acid for electron spin resonance studies (LANDSBERGER et al., 1971, 1973; KORNBERG and McCONNELL, 1971), or perylene for fluorescence polarization studies (RUDY and GITLER, 1972), and compare each of these with their behaviour either in solution or in a known (or presumed) lipid bilayer structure. The results for influenza virus have been interpreted as showing that the viral lipid is in the form of a fluid bilayer when the reference probe is incorporated into a red blood cell membrane (LANDSBERGER et al., 1971). JOST et al. (1973) have concluded, from ESR studies by this method on reconstituted lipoprotein membranes, that a monomolecular layer of immobilized lipid surrounds membrane proteins, while the remainder of the lipid behaves as a fluid bilayer. This indicates a significant drawback of the spin probe method when applied to intact virus: the probe must be added and allowed to be incorporated into the envelope solely by diffusion, whereas reconstitution of membranes as carried out by JOST et al. (1973) can be performed in the presence of the label. If there is specific interaction between envelope structural proteins and at least a small proportion of the lipid, as indicated by the results of TIFFANY and BLOUGH (1969a, b) for influenza and Newcastle disease viruses, the probe molecule will not diffuse as readily into this tightly-bound region as into loosely-bound or non-specific lipid regions. Thus the ESR spectra will not indicate any great degree of interaction between protein and lipid, and will tend to imply that all the viral lipid is in a fluid state. LESSLAUER et al. (1972) have pointed out that the bulky nitroxy spin label group may in fact alter the molecular architecture of the membrane, and prevent ready substitution of labelled for unlabelled molecules. The "melting" behaviour of the lipids, and hence their apparent fluidity, may be affected by the presence of these bulky groups, as pointed out by Hubbell et al. (1970) in relation to the original work of Hubbell and McConnell (1968). Fluorescent electron transfer techniques (Wu and Stryer, 1972) show a great deal of promise in measuring the distance of lipid polar head groups from membrane structural proteins; however these techniques have not yet been applied to viral systems. Unfortunately, many of these probe techniques are limited by the inability of the extrinsic probe to detect subtle changes in membrane structural polypeptides.

B. Information from Disrupted Virions

As with intact virus, electron microscopy is frequently the first line of approach, to determine what effect disrupting agents have had on the particle, and what shapes are adopted by released subunits. Freeze-fracture methods appear to reveal some of the polypeptides located in or around apolar regions of the envelope (Bächi et al., 1969; Nermut and Frank, 1971; Brown et al., 1972; Bächi and Howe, 1973), since the plane of fracture follows these regions (Deamer and Branton, 1967); no details of organisation of the lipid can be seen, since although Branton (1969) has shown recognisable differences to exist between the cleavage planes of lamellar and hexagonal phospholipid phases, an insufficient extent of lipid is revealed within a virus envelope by cleavage.

The isolated substructures obtained depend largely on the methods of disruption and the agents used. For purposes of revealing polypeptides, the most widely-used are detergents and lipid solvents such as Tween 20, sodium dodecyl sulphate (SDS), Nonidet P40, sodium deoxycholate and diethyl ether. Generally these techniques involve extraction of lipid and stabilization of remaining exposed hydrophobic regions of protein; an exception is proteolysis using enzymes such as bromelain where the aim is selective destruction as a means of identification of the sites of constituent proteins (Compans et al., 1970). Once disrupted, the released polypeptides are generally resolved by polyacrylamide gel electrophoresis; this technique has largely superseded earlier methods such as cellulose acetate strip electrophoresis (Laver, 1964) which in some cases failed to resolve viral polypeptides adequately. It is of great use in finding the number of polypeptides (Compans et al., 1970; Haslam et al., 1970; Skehel and Schild, 1971; Lazarowitz et al., 1971) and in determining the degree of contamination by host cell polypeptides (Spear and Roizman, 1972; Holland and Kiehn, 1970). Detergent disruption may be followed by alkylation and reduction before separating the products on polyacrylamide gels, and by appropriate pulse-chase studies the times of synthesis of virus-specific materials in the infected cell can be established. Molecular weights are frequently quoted, using appropriate protein standards, but it must be borne in mind that the proportion of carbohydrates in glycoproteins may materially affect the results (Segrest et al., 1971). Alternative methods of separation of isolated polypeptides include affinity columns (e.g. phytohemagglutinin linked to Sepharose (HAYMAN et al., 1973).

Extensions of the technique of BRETSCHER (1971) to label accessible ε-amino groups and amino-containing phospholipids (GAHMBERG et al., 1972a), or lactoperoxidase with ^{125}I to label surface-accessible tyrosine residues (PHILLIPS and MORRISON, 1971) have been applied to viruses. Polyacrylamide gel electrophoresis of solubilized polypeptides then distinguishes between external (envelope) and internal polypeptides, or even between those on the envelope surface and those in its interior (STANLEY and HASLAM, 1971; KATZ and MARGALITH, 1973). If done in conjunction with radioisotope labelling of monosaccharides, these methods show all the envelope glycoproteins to be external (CARTWRIGHT et al., 1970; COMPANS et al., 1970; KLENK et al., 1972). Enzymatic localization of envelope glycoproteins has also been done using chymotrypsin (SCHULZE, 1970), bromelain (COMPANS et al., 1970) and caseinase (REGINSTER and CALBERG-BACQ, 1968).

Additional techniques include fluorescent labelling, e.g., dansylation of both polypeptides and hexosamines (BOLOGNESI et al., 1973), or the use of lectins to locate glycolipids (OKADA and KIM, 1972; KLENK et al., 1972) as well as specific glycosidases (BIKEL and KNIGHT, 1972). The use of pure phospholipases to localize structures within the viral envelope has not yet been applied to the same extent as for red blood cell ghosts (VERKLEIJ et al., 1973; ZWAAL et al., 1971), although preliminary studies of the action of purified phospholipases on the envelope of Semliki Forest virus show evidence of asymmetry of distribution of lipids between interior and exterior of the envelope (BLOUGH and RENKONEN, unpublished data).

Few amino acid compositions of virus envelope structural polypeptides have so far been determined; this work is limited to a large extent by the availability of material. Techniques are now being developed to permit production of larger quantities of viral components (STANLEY et al., 1973; GREGORIADES, 1973). Earlier studies on the tryptic digests of viral polypeptides are vitiated by the fact that many of these "maps" were done on impure preparations. However "fingerprints" of purified nucleocapsid and envelope polypeptides have been obtained for Semliki Forest virus (SIMONS et al., 1973).

The way in which lipid and protein are embedded in the envelope has not been resolved satisfactorily by electron microscopy. Attempts to digest the viral envelope suffer from the drawback that one may at the same time be damaging structural portions of the envelope. This has of course in many cases been controlled by limiting proteolysis times and/or concentrations (e.g. COMPANS et al., 1970). The structure of the envelope following proteolytic digestion may however be quite different from that of the native envelope, due to rearrangement of the components as balancing forces are disturbed, although spin resonance studies on intact but spikeless influenza virus particles fail to reveal such changes (LANDSBERGER et al., 1973). It has been shown that tryptic digestion of red blood cell ghosts causes an aggregation of the 70 Å intramembranous particles (PINTO DA SILVA and BRANTON, 1970).

C. Information from Reconstituted Viral Membrane Systems

Direct studies on the interaction of viral lipids and proteins or glyco-proteins have been limited thus far to recombinations of the components to form rather ill-defined vesicles, which however do indicate specificity of lipid-protein binding and exhibit similar envelope surface properties to the intact virus (Hosaka and Shimizu, 1972a, b). Exactly how these moieties may combine to form a viral envelope is still not entirely clear; we shall discuss some of the remaining conceptual difficulties in Section IV. A major drawback to the interpretation of structure from such reconstitutions is that, while undoubtedly complex vesicular structures are produced, they commonly exhibit the same surface features on both inner and outer faces, whereas these features are shown on only one face of the intact virus envelope. It scarcely seems possible, considering the data available on asymmetry of distribution of both glycoproteins and lipids, that the same lateral cohesive forces can operate in a reconstituted vesicular membrane showing the same type of polarity on both sides. It is impossible, in this type of experiment, to control the distribution of phospholipids so that the outer layer of a lipid bilayer shall predictably contain a specified excess of phosphatidylcholine or deficit of phosphatidylethanolamine over the inner layer (Thompson and Sears, 1974; Israelachvili, 1973; Michaelson et al., 1973). Obviously, probe and other studies (ESR, NMR, etc.) will have to be done to monitor and evaluate these reconstitution studies; promising techniques using asymmetric phospholipid vesicles are also being developed which may surmount some of the technical problems involved (Thompson and Sears, 1974).

Specific binding between isolated lipids and proteins has not been examined with the same thoroughness for viruses as for mitochondria (Green and Perdue, 1966), chloroplasts (Ji and Benson, 1968) or high density lipoproteins (Scanu and Tardieu, 1971); however, Gregoriades (1973) has isolated a membrane structural polypeptide from influenza virus and determined its amino acid composition (which includes about 10% of amino acids with hydrophobic sidechains), and this will lend itself to detailed lipid binding studies. Similar preparations and analyses have also been done by Laver and Baker (1972) and on the paramyxovirus SV5, which has as many as 64% hydrophobic amino acids (McSharry et al., 1972).

III. Biosynthesis of Envelope Components

A. Lipids

Since we have already discussed the biosynthesis of viral envelope lipids in some detail in our previous review (Blough and Tiffany, 1973), we shall only present a few selected points here. Lipid must be considered a major structural component of enveloped viruses, since in most cases it makes up 20–35% of the weight of the particle; there are some obvious exceptions to this, such as vaccinia virus, which contains only 5% lipid. Until recently, lipids have been thought to be entirely preformed, i.e. derived from lipids already present in the host cell at the time of infection (KATES et al., 1961; PFEFFERKORN and HUNTER, 1963). In cells infected with influenza or Sindbis virus, the specific activity of ^{32}P-labelled phospholipids following maximal growth of the virus was the same for both the virus and the host cell membranes. However, ^{32}P-orthophosphate is a poor label for measurement of turnover in eukaryotic cells, since it becomes compartmentalized in slowly turning over organic pools, making measured turnover rates unreliable (WEINSTEIN and BLOUGH, in press).

A more reliable method of measuring rapid turnover is by pulse-chase studies using 2-^{14}C- and 2-^{3}H-glycerol (BLOUGH and WEINSTEIN, 1973; BLOUGH et al., 1973; GALLAHER et al., 1973 b; GALLAHER and BLOUGH, in press); these studies show that the half-life of major phospholipids in the cell is 2–3 hours, i.e. considerably less than the single-cycle growth time of mammalian viruses. Also, in contrast to earlier studies, when chick embryo fibroblasts were labelled with 2-^{14}C-glycerol, infected with influenza virus (strain A_0/WSN) and then pulsed with 2-^{3}H-glycerol, 30–70% of the viral lipid was found to contain ^{3}H-glycerol, suggesting that newly-formed lipids as well as preformed lipids are incorporated into the virion (BLOUGH, 1974). The presence of unique sequences of hydrophobic amino acids of membrane or envelope structural polypeptides would provide a possible method for preferential selection of lipids bearing certain acyl chains. Alternatively, strong polarity such as that shown by the envelope structural polypeptide of PM2 virus, which has an isoelectric point of pH 12.3 and hence is normally strongly positively charged (SCHAEFER et al., 1974), could select phospholipids on the basis of charge. Thus the so-called "hot spots" where envelope biogenesis is occurring (BEN-PORAT and KAPLAN, 1972) may be sites where newly-synthesized lipid is incorporated into membranes to produce virus of high specific activity. It appears that assembly of the envelope requires a coordinated synthesis and breakdown of lipids (BLOUGH et al., 1973). If turnover is decreased, or if there is lack of coordination between synthesis and turnover of phospholipids or neutral lipids, an abortive or an incomplete infection may occur. In support of this hypothesis, it was found that turnover rates for phospholipids were greatly increased in HeLa cells persistently infected with mumps virus (BLOUGH, 1973); in this case neutral lipids were unaffected. In addition, alternative biosynthetic pathways appear to be operative in cells in the carrier state (BLOUGH, unpublished observations). The function of lipids is not clear at this point, but it was suggested by BLOUGH and TIFFANY (1973), and elsewhere in this review, that lipids are important for the transport of certain hydrophobic membrane-type M proteins to the site of assembly of the virus envelope. The function of glycolipids is at present also unknown, although it has been suggested by BLOUGH and LAWSON (1968) that they play an important role in cell fusion.

B. Carbohydrates

The synthesis of oligosaccharides is controlled by a non-template mechanism consisting of host cell glycosyltransferases. These enzymes are located in membranes of the Golgi apparatus and the endoplasmic reticulum and are responsible for initiation, elongation and termination of oligosaccharides attached to nascent polypeptide chains and ceramides (Hagopian et al., 1968; Cacam and Eylar, 1970; Brady and Fishman, 1973; Grimes and Burge, 1971).

These glycosyltransferases appear to be altered in some way following virus infection, in configuration and/or specificity, since different sequences of monosaccharides are added to viral polypeptides and glycolipids in comparison to those found in normal uninfected plasma membranes or endoplasmic reticulum (Froger and Louisot, 1972a, b; Defrene and Louisot, 1973). Increased activities of mannosyl and N-acetyl-glucosaminyltransferases have been observed in cells infected with Sindbis virus or myxoviruses in comparison to uninfected controls. The K_m of the enzymes was unchanged, although minor changes were observed in pH optima, suggesting that two slightly different host cell glycosyltransferases were involved following infection with Sindbis virus (Grimes and Burge, 1971). Using exogenous acceptors and measuring CMP-sialyl- and fucosyltransferase activities, no differences were noted in specific activity or acceptor specificity. Grimes and Burge (1971) concluded from these results that the carbohydrate portion of viral glycopeptides is therefore host-specified. This is undoubtedly true for many of the smaller viruses, but may not be so for larger viruses such as herpes simplex virus, which has a genomic molecular weight of about 100 million and codes for 47 polypeptides (Heine and Roizman, 1973). One or more of these may in fact be a glycosyltransferase. Ray and Blough (1974) have recently shown that the glucosyltransferase activity of herpes simplex virus (HSV-1)-infected cells is markedly enhanced during infection, while the mannosyltransferase activity is unchanged.

By inhibiting glycosylation of viral polypeptides and glycolipids by using sugar analogues such as 2-deoxy-D-glucose, the sequence of incorporated monosaccharides may be altered and the oligosaccharide chain terminated (Kaluza et al., 1972; Gallaher et al., 1973a; Ray and Blough, 1974). Such sugar analogues offer a convenient means of determining structure and function in component parts of the virion (Courtney et al., 1973).

The major function of the sugar moiety seems to be to ensure the transport of glycoproteins from one membrane (i.e. the site of synthesis) to the final site of assembly or function—the plasma membrane for myxoviruses, or the nuclear envelope for herpesviruses. Support for this hypothesis has recently been provided at the membrane level by Melchers (1973) for monomeric IgG$_1$ immunoglobulin, where the addition of intermediate sugars such as galactose is necessary for transport from the rough to the smooth endoplasmic reticulum. Completed oligosaccharide chains are also necessary for full expression of a viral function such as attachment (BIKEL and KNIGHT, 1972). Sugar molecules may confer the necessary tertiary or quaternary structure on envelope structural polypeptides as well as permit function of non-structural polypeptides such as the glycoprotein responsible for cell fusion by herpesvirus (LEVITAN and BLOUGH, in the press).

It is not clear from any published work on the carbohydrate composition of the virion or of infected or transformed membranes (MORA et al., 1971) exactly how the sequence of sugars is altered from that of normal cells. Detailed information on all aspects of glycosylation in the normal and infected cell will be necessary if the role of glycoproteins and glycolipids in viral infection is to be fully understood. An additional factor not mentioned above is post-synthetic modification of carbohydrate by either endogenous or exogenous glycosidases, to produce glycoproteins and glycolipids with monosaccharide sequences different from those originally synthesized. The only evidence of this at present is in the absence of sialic acid from the envelope hemagglutinin and glycolipids of myxoviruses (KLENK and CHOPPIN, 1970; KLENK et al., 1970).

C. Proteins

Viral structural proteins are synthesized on cytoplasmic ribosomes and must then migrate to the parts of the cell where virus assembly takes place. Appropriate pulse-chase studies have shown that the movement of viral polypeptides from smooth endoplasmic reticulum to plasma membrane is exceptionally rapid, and that with both RNA and DNA viruses (myxoviruses, rhabdoviruses and poxviruses) glycoprotein synthesis is detected as early as one hour after infection (MUDD and SUMMERS, 1970; PRINTZ and WAGNER, 1971).

The messenger RNA of poxviruses is reported to consist of two classes—late messenger RNA with a half-life of 13 minutes and early messenger RNA with a half-life of 120 minutes (SEBRING and SALZMAN, 1967), although others find no difference in the stability of early and late messenger RNAs (ODA and JOKLIK, 1967).

The number of polypeptides incorporated into the virion can be extremely small, as in the case of the alphaviruses; these may have only two envelope proteins, in addition to the nucleocapsid protein (SCHLESINGER et al., 1970; SIMONS et al., 1973). Large viruses such as herpesvirus, on the other hand, code for as many as 47 polypeptides, of which about 25 may be present in the virion (SPEAR and ROIZMAN, 1972). The number of glycosylated polypeptides is also variable, as is the complexity of the carbohydrate moiety. Some simpler viruses contain a large variety of monosaccharides, whereas with the complex poxvirus the two glycoproteins (SAROV and JOKLIK, 1972) contain only the simple sugar N-acetyl-glucosamine (GARON and MOSS, 1971).

Several examples are known of post-translational cleavage of viral polypeptides: proteolytic enzymes are capable of producing the paramyxoviruses SV5 and Sendai with enhanced infectivity, fusion and hemolytic activity (Homma and Tamagawa, 1973; Scheid and Choppin, 1974); the hemagglutinin of influenza virus may be modified by plasmin so that its dimeric structure (Laver, 1971) is cleaved into two large and two small fragments without affecting the biological activity of the virus (Lazarowitz et al., 1973). Additional changes which might influence the charge density of polypeptides include phosphorylation, which has been described for rhabdoviruses (Sokol and Clark, 1973) and murine leukemia viruses (Strand and August, 1971), among others.

IV. The Assembly Process — Theoretical Considerations

We shall consider in this section some of the still unresolved problems involved in assembly of enveloped viruses. Since information is not complete for any group of viruses, we must deal with the subject in rather general terms; some reference to specific virus examples is included, but we shall cover certain aspects of particular groups in more detail in Section V. There is essentially no difference in principle between acquisition of the envelope and release by budding from the plasma membrane, and assembly either by budding into cytoplasmic vesicles which then void following fusion with the plasma membrane, or by budding from the nucleus through the inner nuclear membrane (herpesviruses), so we shall refer only to "the membrane" as the site of assembly. Poxviruses generally assemble by a *de novo* aggregation within the cytoplasm, but an envelope may also be acquired by budding through the plasma membrane. It has been suggested that only the fully enveloped particles are infectious. This group is also more resistant to disruption by organic solvents than other enveloped viruses, perhaps indicating a different membrane structure. The remarks here on envelope assembly may therefore not be applicable to the poxviruses.

A. Production of Materials

Summarising what has been mentioned above under biosynthesis, virally-coded polypeptides may be produced on polyribosomes either in the required size or as part of a larger precursor which must be cleaved within the cell before incorporation into a functional envelope can take place (Klenk and Rott, 1973; Katz and Moss, 1970). Further modification may take place after release, but without obvious structural requirement (Lazarowitz et al., 1973). Post-translational modification also includes glycosylation to form glycoproteins and glycolipids (and may involve elaboration and then shortening by endogenous glycosidases) as a sequential procedure determined by host cell enzymes.

Lipid metabolism of the cell is considerably altered following viral infection, to the extent that a large proportion of viral lipids are newly-synthesized, and are not simply drawn from pre-existing pools of normal cell membrane lipids. Otherwise the same classes of lipid are generally found in the envelope as in the parent membrane, although the proportions of polar classes and the distribution of acyl chains within each class may vary (BLOUGH and TIFFANY, 1973).

B. Transport to the Assembly Site

Once synthesized, viral components must be transported to the actual site of assembly. This may present no difficulty in the case of lipids and some polypeptides, but structural proteins having sizable non-polar regions used in hydrophobic bonding require some form of protection in transit to preserve their unique tertiary structure. This could take the form of specific binding of lipids to form a soluble lipoprotein complex, or of a detergent-like action of lipid in forming a "hemi-micelle" surrounding the non-polar regions of the protein. In either of these cases a means would also be offered for introduction of specifically-tailored lipids into the assembly region. In this sense the viral polypeptide could act like the lipid carrier protein of uninfected cells (WIRTZ and ZILVERSMIT, 1968).

C. Insertion into the Parent Membrane

The sequence of insertion of components into the membrane, and conversion of an area of membrane into a budding patch, may be determined by the type of virus to be formed. Viruses formed by budding can be divided into (a) those having a loose envelope of rather variable shape which seems not to be tightly attached to the underlying nucleocapsid core (e.g. influenza, herpesviruses), and (b) those where the envelope is tightly packed onto the core (e.g. arboviruses, and rhabdoviruses except for the flattened end region of the particle). In group (a) myxoviruses are known to have an envelope structural protein or M protein forming a coherent but somewhat flexible shell within the envelope. This shell has insufficient cohesion in the absence of viral lipid to be insolated as such without preliminary cross-linking with glutaraldehyde (SCHULZE, 1970). The presence of a similar protein or group of proteins has not been demonstrated for herpesviruses, but at least ten structural polypeptides have been identified which are neither glycosylated (envelope outer surface) or core, and hence might fulfil this function (HEINE and ROIZMAN, 1973). In (b), the nucleocapsid core is complete and tightly packed before budding begins (e.g. Semliki Forest virus, GRIMLEY and FRIED-MAN, 1970). The same is true of the rhabdoviruses except that ordering of the nucleocapsid into a tight cylindrical spiral takes place during rather than before budding (HOWATSON, 1970). Rhabdoviruses are thought to contain envelope M protein (KANG and PREVEC, 1970), and specific binding may take place between envelope proteins and nucleocapsid protein, on the basis of numerical correspondence between these polypeptide species (SOKOL et al., 1971; NEU-

RATH et al., 1972). This seems to be indicated also by phenotypic mixing experiments involving double infection with vesicular stomatitis virus (VSV) and paramyxovirus SV5; particles contained surface glycoproteins from either virus, but VSV nucleoprotein was associated only with VSV membrane structural protein (McSHARRY et al., 1971). However, the rhabdovirus envelope appears to have integrity quite distinct from that conferred by nucleocapsid binding, in the flattened base region where a ballooning or re-entrant form may be shown. Thus the important factor in rhabdovirus envelope assembly seems to be the time sequence of coiling of nucleocapsid, and this could conceivably be under the influence of a very small number of molecules of "morphopoietic factor". The critical requirement for both (a) and (b) is for a structural component capable of forming a base on which the envelope lipids and surface polypeptides can be assembled. It seems likely also that the characteristic curvature of the envelope both in budding and in the mature particle will be determined by this structural factor, perhaps also aided in part by lateral repulsion between external projections of the virion.

It is not known exactly how proteins such as the glycosylated surface proteins of the virus are inserted through the lipid region of the host membrane, nor how, once inserted, they are anchored in position. We have suggested a possible method, based on a structural model for influenza virus (BLOUGH, 1969; TIFFANY and BLOUGH, 1970a), involving local phase changes of the membrane lipid region from a bilayer to a micellar form, which would permit passage of the external proteins and leave them anchored to M protein bases with the lipid remaining in micellar form (Fig. 1A). If the lipid of the envelope is, in fact, in bilayer form, as is suggested from spin resonance studies of influenza virus and X-ray diffraction analysis of Sindbis virus, rearrangement of lipid from micellar to bilayer form could follow the insertion. Calculations by OHKI and AONO (1970) indicate that bilayer-hexagonal lipid phase rearrangements may take place within the range of net charge of 0–2e per phospholipid molecule, with the bilayer being the lowest energy form. Higher net charge densities at the polar end of envelope surface proteins may aid in the preliminary conversion, and the passage into the membrane of the apolar inner end of the surface protein might then reduce local net charge sufficiently to induce reversion to a bilayer configuration of lipid. In the proposed influenza model, subsequent surface proteins were inserted alongside the first with lateral interactions between M protein bases holding the units together until a patch of cell membrane was entirely infiltrated by viral material (Fig. 1B). Prior insertion of such units might facilitate addition of further units, but this is not strictly necessary to the model. Conceivably a mechanism of this type could operate in the normal insertion of glycosylated cell membrane proteins which then show lateral mobility (SINGER and NICOLSON, 1972), as an alternative to the reverse-pinocytosis method proposed by HIRANO et al. (1972). A similar process of fusion between virus-specific cytoplasmic vesicles and the plasma membrane has been suggested for myxovirus assembly (CHOPPIN et al., 1972; Fig. 1C). If lateral mobility exists, then the point of insertion

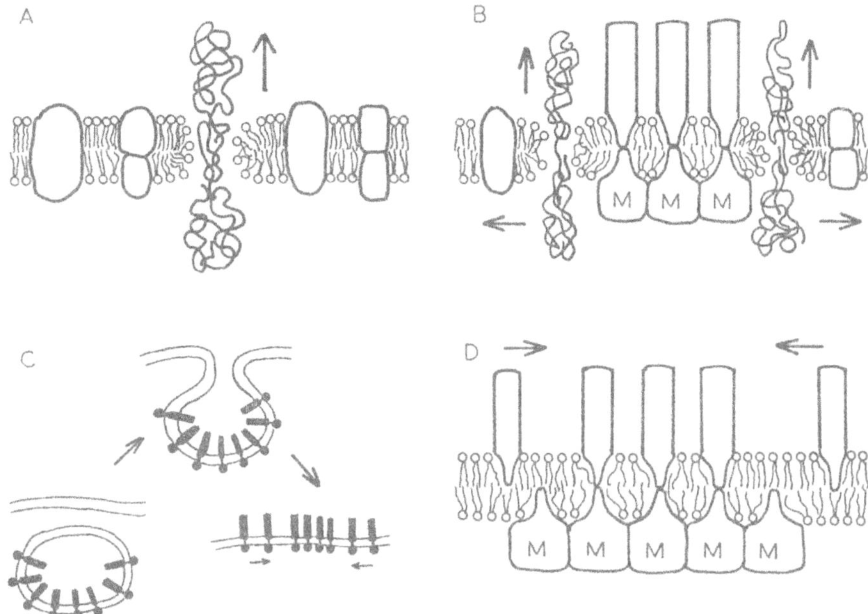

Fig. 1 A–D. Models of assembly of virus-specific areas of membrane. Note that a complete penetration of the membrane by spike and M protein is not an essential part of any of these models. The models are not meant to indicate one-to-one correspondence between spikes and M protein units. (A) Insertion of spike protein (shown here already associated with M protein) through the membrane. A local phase change of membrane lipid aids insertion. The membrane structure shown is the fluid mosaic model of SINGER and NICOLSON (1972). (B) Presence of an inserted spike makes further insertion favorable, with a similar lipid phase change in each case. In the completed region the lipid is shown to have reverted to a bilayer configuration. Membrane proteins are gradually displaced laterally since a coherent "raft" of viral M protein is formed. (C) Method of insertion by fusion of virus-specific vesicles with the membrane as suggested by CHOPPIN et al. (1972). (D) Model showing lateral mobility of spikes and M protein. Lateral cohesion of M proteins prevents redispersal

of viral glycoproteins may not, in fact, be at the subsequent site of budding (Fig. 1 D). If more random entry of surface proteins is postulated, budding itself must be preceded by the presence at the inner membrane surface of the major envelope structural factor—a firm nucleocapsid "former" or a raft or island of structural M protein.

D. Organisation of the Envelope Prior to Budding

If insertion of envelope external proteins is essentially a random process, lateral diffusion of surface units, floating in a sea of membrane lipid, would take place until the area over the "former" contained a full complement of surface proteins (not the full amount required for the envelope, but enough to initiate budding). A major problem at this point, however, is the mechanism of recognition across the lipid region between surface proteins and "former". It has been generally assumed, in envelope models postulating a lipid bilayer (Choppin et al., 1972; Klenk, 1973), that surface proteins of the envelope have a hydrophobic end which is inserted into the outer leaflet of the bilayer in the same manner as in the fluid mosaic membrane model (Singer and Nicholson, 1972); this is to some extent supported by evidence on the manner of aggregation of isolated envelope subunits (Laver and Valentine, 1969) and from the extensive non-polar amino acid composition of Semliki Forest virus glycoprotein (Gahmberg et al., 1972b). No evidence exists at the moment for a corresponding insertion of "former" protein from the inner side of the bilayer (Lenard et al., 1974), but it is hard to see how laterally-mobile surface proteins could be constrained to aggregate in a patch overlying the "former" unless some direct interaction through the bilayer anchors them into position (Fig. 2A). This anchoring is an essential part of our model for the influenza virus envelope (Tiffany and Blough, 1970a), although we are prepared to concede that the lipid region may be substantially in the form of a bilayer. In addition, the fluidity indicated to exist in the lipid region of the influenza virus envelope (Landsberger et al., 1971, 1973) from spin-resonance studies would tend to make interaction between M protein and surface proteins across an intact inner bilayer leaflet even more difficult. The relative numbers of spike and envelope polypeptides calculated for rabies (Neurath et al., 1972; Sokol et al., 1971) and vesicular stomatitis viruses (Cartwright et al., 1972), suggest that surface protein units may be anchored to the "former" in the rhabdovirus envelope. Unless such an anchoring takes place, it is hard to explain recognition between external and internal proteins of the envelope, sufficient to maintain cohesion of the "raft" without loss by lateral diffusion, and to displace host membrane proteins from the area, since none are found in mature virus (Holland and Kiehn, 1970). Fig. 2B shows a possible method, which involves specific interaction between inner bilayer leaflet lipids and the M protein; lipid fluidity would be limited largely to the outer leaflet in this model.

E. Bud Growth and Aberrant Forms of the Envelope

At the time of initiation of budding, it is unlikely that all the envelope materials required have already entered the membrane. Subsequent enlargement of the bud must then take place by lateral diffusion and/or direct insertion into a "growing ring" surrounding the line of attachment of the bud to the membrane. It is in this stage of assembly that anomalies of envelope structure become apparent for myxoviruses grown in the presence of detergents (Blough, 1963a) or exogenous lipids (Blough, 1963b, 1964) or high titre passage "incomplete" or von Magnus virus (Blough et al., 1969; Blough and Merlie, 1970). These forms of virus are markedly pleomorphic, and strains which are normally spherical frequently show filamentous particles. Exogenous branched chain fatty acids have been shown to be incorporated into phospholipids of influenza virus, thus changing the composition of viral lipids (BLOUGH and TIFFANY, 1969), and the lipid metabolism of the host cell is considerably altered during production of von Magnus virus (BLOUGH and WEINSTEIN, 1973). Several mechanisms can be postulated for this change in shape:

a) The presence of exogenous lipids or shifts in availability of cellular lipids causes "expansion" of the lipid region and weakening of forces within the envelope, thus making it more flexible.

b) Selective binding of certain lipids to M protein produces two types corresponding roughly to "penton" and "hexon" of regular viruses, in which relatively few "pentons" are available, leaving predominantly "hexon", which tends to form tubular filaments of indeterminate length.

c) There may actually be two closely-related M proteins of very similar molecular weight and physical properties, which have a hexon-like or penton-like function. The relative availability of these may depend on host type and growth conditions.

d) Modification of normal viral lipid composition may be less important than the time sequence of changes in infected cell metabolism leading to restrictions in the proportions of envelope materials available at the budding site (possibly through influencing lipid-aided transport of proteins). Thus although enough "penton" material may be available to initiate budding, the local supply may fall off so that less is subsequently available during the rounding-up and pinching off of the virus bud, and the resultant local excess of "hexon" produces filaments.

Normally filamentous strains of influenza (e.g. influenza C) or SV5 may be produced in a similar manner, but here the nucleocapsid may act to a greater extent as a "former" since it appears to be uniformly helically wound in a continuous strand against the inner wall of the envelope, unlike the loosely-bundled nucleocapsid of influenza A and B.

F. Release

Bud termination in loosely-enveloped viruses may also be determined by local availability of "penton" material, resulting in a sudden necking down of the envelope bud. Little is known of the mechanism of pinching off and resealing of membrane and envelope at this time, but the process presumably resembles that involved in pinching off membrane fragments (such as the "normal cell particles" released from the allantois of embryonated hen's eggs) or in intracytoplasmic vesicle formation; it has in fact been suggested that virus release is only a modification of this normal exfoliation phenomenon (KINGSBURY, 1972). After pinching off, release of the particle from the cell (assuming it buds at the cell membrane) may be influenced by accessibility of receptor groups for the virus on the surrounding cell surface. Influenza virus is pinched off but remains bound at the cell surface if antibody directed against viral neuraminidase is present (SETO and CHANG, 1969; DOWDLE et al.,

1974). Pretreatment of the cell during the eclipse phase with the antibody tends also to produce filamentous rather than spherical particles. Conceivably an effect similar to that in Sec. IV E (b) or (c) might operate, in which neuraminidase was preferentially bound to the "penton" rather than the "hexon" component; the presence of antibody to neuraminidase would than sequester the enzyme and cause "hexon" and hence filamentous structures to predominate.

G. Forces Operating during Assembly and in Maintenance of Structure

Until recently, relatively little attention has been given to the types of interaction involved in assembly and maintenance of the integrity of the viral envelope. It is common practice to use the terms hydrophobic and hydrophilic loosely to justify a particular arrangement of molecules within a particle, but this may not be sufficient to determine whether the postulated structure will be stable, or whether a logical series of assembly steps can take place to produce this structure.

The information at present available on the types of forces actively contributing to envelope stability comes partly from experiments with various disruptive agents, and partly by an extension of the available information on cell membrane structural factors to the case of the viral envelope. Thus by a judicious choice of agent (detergents, lipid solvents, proteases or lipases) or changes in pH or ionic strength (Waite et al., 1972), virus particles may be partly dismantled, and the ease with which this is done gives some indication of the milieu of subunit molecules prior to release. Another approach involves mixing isolated viral components (e.g. lipids and proteins) in an attempt to reconstruct the particle under controlled conditions, but this is subject to experimental difficulties as mentioned above.

One should also perhaps be wary of attempts to predict forces or structural arrangements in the viral envelope from data on the organisation of the parent membrane at which the virus was assembled. Although the trilamellar staining pattern of the parent membrane seems to continue into the virus envelope during budding, host proteins are not present in the budding region (Holland and Kiehn, 1970; Aoki et al., 1970), and a quite different pattern of interaction between protein and lipid may exist in the virus envelope from that in the unperturbed membrane without change in the stained electron-optical appearance. Since there is no synthesis or turnover of envelope components in the mature virus and its necessary functions are few, it seems likely that the fluidity and fluctuating composition which are desirable in a functional membrane will be unnecessary or possibly structurally undesirable in the virus. In addition, the procedures used in preparation of specimens for electron microscopy may themselves introduce an apparent similarity of structure.

Nevertheless, the same general principles must apply within virus envelopes as in other structures involving protein and lipid molecules, since similar polar, ionic and non-polar chemical groups are present in both cases. The possible magnitude of these interacting forces can be assessed, although of course many variations can occur in an actual viral system through minor changes in composition, etc.

The effects likely to be of importance are electrostatic or ionic forces, permenent and induced dipole interactions, interpeptide or sidechain hydrogen bonds in proteins, and dispersion or London-van der Waals forces. The magnitudes of some of these have been estimated by SALEM (1962a, b): electrostatic interaction energies (attractive or repulsive) are in the region of 5 kcal/mol for a pair of univalent charged groups 5 Å apart, depending in part on the effective (microscopic) dielectric constant of the medium between them; induced dipole energies (e.g. between an ionic or polar group and a dipole induced in a —CH_2 group) are typically less than 0.1 kcal/mol and can probably be ignored; dispersion forces are attractive but fall off very rapidly ($\infty\ 1/D^6$) with increasing separation D of the interacting groups, and energies between two groups are small (ca. 0.1 kcal/mol), but if summed over all the groups of closely-apposed parallel hydrocarbon chains the resultant may be as high as 8 kcal/mol for a pair of stearoyl chains 5 Å apart in a monomolecular film. Similar magnitudes are also possible between strong permanent dipole groups. Van der Waals forces are not limited to hydrocarbons, but can also operate between saturated or largely apolar regions of proteins, and between proteins and lipids. These forces are of particular importance in extensive and closely-packed water-insoluble regions such as the hydrocarbon tail region of a phospholipid bilayer, where the net ionic interaction is either near zero or repulsive. Because of their extreme sensitivity to separation of the interacting groups, Salem has termed these summed dispersion forces "distance-sensitive" (SALEM, 1962a, b). These forces may hence be of comparable magnitude to ionic forces; however, the hydrogen-bonding nature of water itself may disturb these interactions (KAUZMANN, 1959) and entropic factors due to limitation of rotation of the hydrogen-bonded groups may influence their effectiveness. Hydrogen bonding of water could be of considerable importance in retention or loss of tertiary structure when a structural protein is extracted from the virus and purified for use in reconstitution experiments. It should be noted that the additivity of ionic forces operates in the opposite direction to that of van der Waals forces even when arrays of positive and negative charges are present; the net interaction energy of such on ionic array is always at least slightly repulsive, whereas van der Waals forces are always attractive.

The contribution of each of the types of force mentioned above to envelope structure can be at least partially estimated from the response to disruptive agents. Ionic forces, especially those depending on net charge neutralization with divalent cations, will be disrupted if the ionic strength of the medium is sharply raised. Hydrogen bonding will be weakened in the presence of high concentrations of a hydrogen-bonding solute such as urea, and disulfide bonds in the same way with sulfhydryl reagents. Lipid solvents and non-ionic detergents, by their ability to generate dispersion forces with non-polar regions of the envelope, will weaken "hydrophobic" structural relationships. Anionic or cationic detergents (such as sodium dodecyl sulfate or sodium deoxycholate) have a similar effect, but are frequently more dependent on other components of the system which may influence their micellar behaviour. These effects may be reversible on withdrawal of the disruptive agent from an isolated fraction of the envelope.

H. The Role of the M Protein in Envelope Structure

The detection and isolation of a virus envelope internal structural protein (the M protein) is one of the most important advances of recent years in the study of enveloped viruses. Such a protein had been predicted for influenza virus (BLOUGH, 1969; TIFFANY and BLOUGH, 1970b) on the basis of the then known dimensions and amounts of protein in the virion. Thus it was known that the envelope "unit membrane" appeared thicker on the inner side, depending on the electron microscopic staining procedure used (COMPANS and DIMMOCK, 1969), and after prolonged protease digestion (KENDAL et al., 1969). Further evidence soon followed (COMPANS et al., 1970) of a protein identifiable on polyacrylamide gels as an envelope component but not one of the external glycoproteins. Since then, envelope structural proteins have also been identified for rhabdoviruses, and classified (WAGNER et al., 1972), and similar proteins will in due course be discovered for other virus groups. Because of its site in the envelope, the M protein must inevitably be considered in relation to the viral lipid, and a number of points remain unresolved with regard to the packing and interrelationship of the two.

TIFFANY and BLOUGH (1970a) gave calculations indicating that a considerable excess of lipid appeared to be present in the influenza virion over that necessary to form a bilayer (80–100%), and the actual volume available in the bilayer would be further reduced by penetration of surface proteins in the same manner as in SINGER and NICOLSON's model (1972) for the cell membrane, as is indicated for many groups of viruses (SCHULZE, 1973; GAHMBERG et al., 1972b; VERNON et al., 1972). The problem is even more acute in the case of leukoviruses, which have as much as 35% lipid, although here the possibility of lipid in the particle core remains unresolved. It is of course hazardous to perform calculations with a virus as notoriously variable in size and shape as influenza, as can be seen on recalculating the particle size necessary to accommodate the lipid content of the A_0/PR_8 strain (average diameter 1000 Å); thus there is 80% more bilayer area in a particle of 1270 Å overall diameter than in a 1000 Å particle (mean bilayer diameter increasing from 760 Å to 1030 Å). It seems clear that the effect of various methods of preparation for electron microscopy on the dimensions of the virion requires closer study (NERMUT and FRANK, 1971). In addition, as mentioned in BLOUGH and TIFFANY (1973), the calculations depend closely on the value taken for the molecular weight of the whole virion; LANDSBERGER et al. (1971) calculate that there is just the amount of lipid required for a bilayer by using a particle

MW of 250×10^6, whereas a more reliable figure is 360×10^6, based both on the protein data of REIMER et al. (1966) and the MW of viral RNA (SKEHEL, 1971). Interpretation of the size of substructures of the envelope from electron micrographs is subject to variation also. SCHULZE (1972) quotes 60 Å both for the M protein shell thickness and the bilayer thickness, whereas X-ray data on lipid bilayers show a thickness of 40 Å (WILKINS et al., 1971) and an alternative interpretation of the staining pattern of the M protein region (30–40 Å thick) is given by NERMUT (1972).

The original model of TIFFANY and BLOUGH (1970a) attempted to overcome this excess of lipid by postulating that the M protein formed specific associations with lipids in the same way as the chloroplast lamellar protein studied by JI and BENSON (1960). This was related to the variations in lipid composition seen in different strains of the same virus type (TIFFANY and BLOUGH, 1969a, b; MCSHARRY and WAGNER, 1971; DAVID, 1971; BLOUGH, 1971; BLOUGH et al., 1967). From the degree of variation both in polar groups and in acyl chains, it would seem that the binding of at least a small proportion of the lipid to M protein is indicated, without specifying whether it is a polar or a hydrophobic interaction. Fig. 2B shows a type of polar interaction which could take place with a non-penetrating M protein, to provide a partially-immobilized lipid region in the neighbourhood of the M protein "raft", capable of serving as a recognition site for floating and laterally-mobile spike proteins. We are still inclined to favour the penetration of the inner bilayer leaflet by M protein (Fig. 2A) as offering assembly and structural advantages, and possibly also as explaining the discrepancies of staining patterns noted between SCHULZE (1972) and NERMUT (1972).

It is of interest to see whether the influenza M protein data of GREGORIADES (1973) may be used in the calculation method of TREMAINE and GOLDSACK (1968), to determine the ability of this protein to form a shell by non-polar interaction alone in the same way as capsid proteins of small regular viruses (BANCROFT et al., 1967). This calculation depends on the assumption that all polar amino acids are located on the inner and outer faces of a spherical shell formed of the protein units. We assume also that the units are cylindrical and that their partial specific volume is 0.72 cc/g. The volume of the protein unit is then 30900 Å³ from its MW of 25900 (SCHULZE, 1970), and its polar area from the TREMAINE and GOLDSACK calculation (1968) is 1411 Å². This corresponds to a mean cylindrical diameter of 30 Å and a shell thickness of 44 Å, in good agreement with earlier data. Using REIMER et al. (1966) value of 4.2×10^{-16} g protein per particle and assuming 50% of this is M protein (SCHULZE, 1970), we calculate that there are 4890 M protein units. This is higher than the values of 3300–3700 units quoted by SCHULZE (1972) who however has used a value of 42% for the proportion of M protein, apparently taken from COMPANS et al. (1970). The area of the hydrophobic sides of the units (2πrh) is 4100 Å² or 75% of the total area. CHOTHIA (1974) suggests a hydrophobic interaction energy of 24 cal/mole/Å² for such surfaces, giving $4100 \times 24 = 99$ kcal/mole per M protein unit. This seems to provide a sub-

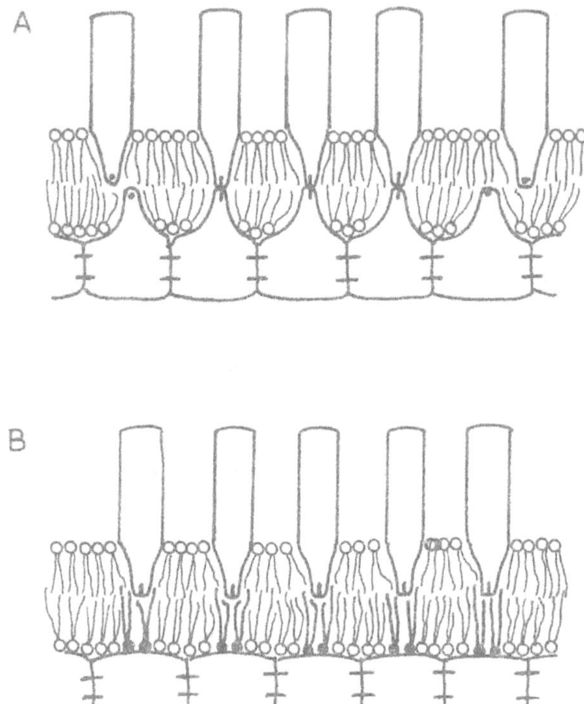

Fig. 2A and B. Penetrating and non-penetrating M protein models. Each M protein unit shown here may consist of about 6 individual molecules. (A) Assembly by lateral diffusion of "floating" spikes into the region where a coherent "raft" of M protein occurs. Spikes remain anchored and the patch remains free of host membrane proteins. Black bars represent hydrophobic interactions. (B) Non-penetrating M protein. The "raft" is coherent, but specific lipid binding in the inner lipid leaflet is necessary to provide a site for recognition by spike proteins. (Note that only a small proportion of viral lipid would be strain-specific in the case, involving only moderate differences between host membrane and viral lipid compositions

stantial stabilizing force within the envelope, and by further interaction between protein and lipid, a robust envelope structure which does not require an internal nucleocapsid former to maintain its integrity (Blough, 1963a).

Rhabdoviruses were considered above to occupy an intermediate position between those viruses whose envelope form is determined by their M protein, and those which depended on a dense core to act as a former during formation of the envelope. From the numerical correspondence of the envelope structural proteins of rabies virus (Sokol et al., 1971; Neurath et al., 1972) and of vesicular stomatitis virus (Cartwright et al., 1972), it appears that there is specific binding between the nucleocapsid protein and an envelope protein, and this has been made the basis for interesting structural models (Vernon et al., 1972; Cartwright et al., 1972). Both of these models show the surface spike glycoproteins penetrating the lipid region, but the nature of the anchoring is not clear. We have calculated from analysis of rabies virus lipids (Blough,

AASLESTAD and TIFFANY, unpublished data) that there is slightly less lipid per particle than is required to form a bilayer, and this may indicate specific binding between spike proteins and envelope protein rather than "floating" spikes. It has been noted that rabies virus is structurally more resistant to ether and detergent treatment than some other enveloped viruses (CRICK and BROWN, 1970), suggesting that lipid plays less part in maintaining the integrity of the envelope. The model of VERNON et al. (1972) accounts for the large hexagons which are apparently a structural feature of some strains of rabies virus examined by negative-contrast electron microscopy, as artifacts produced by superimposition of images from the upper and lower surfaces of the particle and the strong parallel banding of the helical nucleocapsid core. KUWERT et al. (1972) calculated that there were about 580 such hexagons with a centre-to-centre spacing of 100 Å; this appears to be based on mismeasurement of their micrographs, which in fact show a 200 Å spacing and hence about 140 hexagons per particle. The connection between this number and that of the spikes, which has been variously estimated as 790 (BLOUGH and TIFFANY, 1973) and 1072–1453 (VERNON et al., 1972), is not yet clear. The number of surface projections on vesicular stomatitis virus is estimated to be about 500 (CARTWRIGHT et al., 1972).

The exact architecture of the rhabdovirus envelope cannot be worked out with the data currently available, but it would seem that the M protein plays a two-fold role which is not seen (or only partially seen) in the myxoviruses: it acts as a base for the spike glycoproteins on the outer side of the envelope, and it also contributes to the stability of the inner helix of ribonucleoprotein within the particle.

It will be seen in Figs. 1 and 2 that we have indicated an apparent one-to-one correspondence between surface spikes and M protein units in the envelope. The numbers of M protein molecules calculated above and by SCHULZE (1972) are considerably greater than the number of spikes calculated from their surface spacing and the viral diameter, corresponding roughly to 6 molecules per spike unit. We do not consider it entirely improbable that the M protein will form sub-aggregates in this way, and each molecule need contribute only a very small "tail" of hydrophobic character to the penetration of the lipid layer. Such a tail would almost certainly be overlooked in the isolated molecule.

References

AOKI, T., BOYSE, E. A., OLD, D. T., HARVEN, E. DE, HÄMMERLING, U., WOOD, H. A.: G (Gross) and H-2 cell-surface antigens: Location on Gross leukemia cells by electron microscopy with visually labelled antibody. Proc. nat. Acad. Sci. (Wash.) 65, 569–576 (1970)

BÄCHI, T., GERHARD, W., LINDENMANN, J., MÜHLETHALER, K.: Morphogenesis of influenza virus in Ehrlich ascites tumor cells as revealed by thin-sectioning and freeze-etching. J. Virol. 4, 769–776 (1969)

BÄCHI, T., HOWE, C.: Morphogenesis and ultrastructure of respiratory syncytial virus. J. Virol. 12, 1173–1180 (1973)

Bancroft, J. B., Hills, G. J., Markham, R.: A study of the self-assembly process in a small spherical virus. Formation of organized structures from protein subunits *in vitro*. Virology **31**, 354–379 (1967)

Ben-Porat, T., Kaplan, A. S.: Studies on the biogenesis of herpesvirus envelope. Nature (Lond.) **235**, 165–166 (1972)

Bikel, I., Knight, C. A.: Differential action of *Aspergillus* glycosidases on the hemagglutinating and neuraminidase activities of influenza and Newcastle disease viruses. Virology **49**, 326–332 (1972)

Blough, H. A.: The role of the surface state in the morphogenesis of influenza virus filaments. Virology **19**, 112–114 (1963a)

Blough, H. A.: The effect of vitamin A alcohol on the morphology of myxoviruses. I. The production and comparison of artificially produced filamentous virus. Virology **19**, 349–358 (1963b)

Blough, H. A.: Role of the surface state in the development of myxoviruses. In: Ciba Symposium, Cellular Biology of Myxovirus infections, G. W. Wolstenholme and J. Knight (eds.), p. 120–143. London: J. & A. Churchill 1964

Blough, H. A.: Studies on the structure of influenza virus envelope. Bull. Wld. Hlth. Org. **41**, 487–488 (1969)

Blough, H. A.: Fatty acid composition of individual phospholipids of influenza virus. J. gen. Virol. **12**, 317–320 (1971)

Blough, H. A.: The effect of chronic viral infections on the synthesis and turnover of lipids in cultured animal cells. Ninth Intl. Cong. Biochem. P407 (1973)

Blough, H. A.: New lipids incorporated into the membrane of influenza virus. Nature (Lond.) **251**, 333–335 (1974)

Blough, H. A., Gallaher, W. R., Weinstein, D. B.: Viral lipids: Host cell biosynthetic parameters. In: Membrane mediated information, P. W. Kent (ed.), **1**, p. 183–199. Lancaster: Medical & Technical Publishing Co. 1973

Blough, H. A., Lawson, D. E. M.: The lipids of paramyxoviruses: a comparative study of Sendai and Newcastle disease viruses. Virology **36**, 286–292 (1968)

Blough, H. A., Merlie, J. P.: The lipids of incomplete influenza virus. Virology **40**, 685–692 (1970)

Blough, H. A., Merlie, J. P., Tiffany, J. M.: The fatty acid composition of incomplete influenza virus. Biochem. biophys. Res. Commun. **34**, 831–834 (1969)

Blough, H. A., Tiffany, J. M.: Incorporation of branched-chain fatty acids into myxoviruses. Proc. nat. Acad. Sci. (Wash.) **62**, 242–247 (1969)

Blough, H. A., Tiffany, J. M.: Lipids in viruses. Advanc. Lipid Res. **11**, 267–339 (1973)

Blough, H. A., Weinstein, D. B.: Effect of influenza virus infection on lipid metabolism of chick embryo fibroblasts. In: Biology of the fibroblast, E. Kulonen and J. Pikkarainen (eds.), p. 303–308. London: Academic Press 1973

Blough, H. A., Weinstein, D. B., Lawson, D. E. M., Kodicek, E.: The effect of vitamin A on myxoviruses. II. Alterations in the lipids of influenza virus. Virology **33**, 459–466 (1967)

Bolognesi, D. M., Luftig, R., Shaper, J. H.: Localization of RNA Tumor virus polypeptides. I. Isolation of further virus substructures. Virology **56**, 549–564 (1973)

Brady, R. O., Fishman, P. H.: Alterations in membrane glycolipids in tumorigenic virus-transformed cell lines. In: Membrane mediated information, P. W. Kent (ed.), **1**, p. 18–63. Lancaster: Medical & Technical Publishing Co. 1973

Branton, D.: Membrane structure. Ann. Rev. Plant Physiol. **20**, 209–238 (1969)

Bretscher, M. S.: Human erythrocyte membranes: specific labelling of surface proteins. J. molec. Biol. **58**, 775–781 (1971)

Brown, D. T., Waite, M. R. F., Pfefferkorn, E. R.: Morphology and morphogenesis of Sindbis virus as seen with freeze-etching techniques. J. Virol. **10**, 524–536 (1972)

Cacam, J. F., Eylar, E. H.: Glycoprotein biosynthesis: Purification and characterization of a glycoprotein: galactosyl transferase from Ehrlich ascites tumor cell membranes. Arch. Biochem. Biophys. **137**, 315–323 (1970)

Cartwright, B., Smale, C. J., Brown, F., Hull, R.: Model for vesicular stomatitis virus. J. Virol. **10**, 256–260 (1972)

Cartwright, B., Talbot, P., Brown, F.: The proteins of biologically active subunits of vesicular stomatitis virus. J. gen. Virol. **7**, 267–272 (1970)

CHOPPIN, P. W., COMPANS, R. W., SCHEID, A., McSHARRY, J. J., LAZAROWITZ, S. G.: Structure and assembly of viral membranes. In: Membrane research, C. F. Fox (ed.), p. 163–185. London-New York: Academic Press 1972

CHOTHIA, C.: Hydrophobic bonding and accessible surface area in proteins. Nature (Lond.) **248**, 338–339 (1974)

COMPANS, R. W., DIMMOCK, N. J.: An electron microscope study of single-cycle infection of chick embryo fibroblasts by influenza virus. Virology **39**, 499–515 (1969)

COMPANS, R. W., KLENK, H.-D., CALIGUIRI, L. A., CHOPPIN, P. W.: Influenza virus proteins. I. Analysis of polypeptides of the virion and identification of spike glycoproteins. Virology **42**, 880–889 (1970)

COOPER, P. D.: A chemical basis for the classification of animal viruses. Nature (Lond.) **190**, 302–305 (1961)

COURTNEY, R. J., STEINER, S. M., BENYESH-MELNICK, M.: Effect of 2-deoxy-D-glucose on Herpes simplex virus replication. Virology **52**, 447–455 (1973)

CRICK, J., BROWN, F.: In: The biology of large RNA viruses, R. D. BARRY and B. W. J. MAHY (eds.), p. 133. New York: Academic Press 1970

DAVID, A. E.: Lipid composition of Sindbis virus. Virology **46**, 711–720 (1971)

DEAMER, D. W., BRANTON, D.: Fracture planes in an ice-bilayer model membrane system. Science **158**, 655–657 (1967)

DEFRENE, A., LOUISOT, P.: Glycoprotein biosynthesis in animals infected with a myxovirus. II. Study of physicochemical parameters of liver N-acetyl-glucosaminyl transferase. Int. J. Biochem. **4**, 249–258 (1973)

DOWDLE, W. R., DOWNIE, J. C., LAVER, W. G.: Inhibition of virus release by antibodies to surface antigens of influenza viruses. J. Virol. **13**, 269–275 (1974)

FROGER, C., LOUISOT, P.: Comportement des mannosyltransférases microsomiques dans les cellules porteuses d'une infection à Arbovirus. C. R. Acad. Sci. (Paris) **274**, 737–740 (1972a)

FROGER, C., LOUISOT, P.: Glycoprotein synthesis in arbovirus-infected cells. II. Study of microsomic mannosyl transferase activity. Int. J. Biochem. **3**, 613–622 (1972b)

GAHMBERG, C. G., SIMONS, K., RENKONEN, O., KÄÄRIÄINEN, L.: Exposure of proteins and lipids in the Semliki Forest virus membrane. Virology **50**, 259–262 (1972a)

GAHMBERG, C. G., UTERMANN, G., SIMONS, K.: The membrane proteins of Semliki Forest virus have a hydrophobic part attached to the viral membrane. FEBS Letters **28**, 179–182 (1972b)

GALLAHER, W. R., BLOUGH, H. A.: Syntheses and turnover of lipids in monolayer cultures of BHK-21 cells. Arch. Biochem. Biophys. (in press)

GALLAHER, W. R., LEVITAN, D. B., BLOUGH, H. A.: Effect of 2-deoxy-D-glucose on cell fusion induced by Newcastle disease and Herpes simplex viruses. Virology **55**, 193–201 (1973a)

GALLAHER, W. R., WEINSTEIN, D. B., BLOUGH, H. A.: Rapid turnover of principal phospholipids in BHK-21 cells. Biochem. biophys. Res. Commun. **52**, 1252–1256 (1973b)

GARON, C. F., MOSS, B.: Glycoprotein synthesis in cells infected with vaccinia virus. II. A glycoprotein component of the virion. Virology **46**, 233–246 (1971)

GREEN, D. E., PERDUE, J. F.: Membranes as expressions of repeating units. Proc. nat. Acad. Sci. (Wash.) **55**, 1295–1302 (1966)

GREGORIADES, A.: The membrane protein of influenza virus: extraction from virus and infected cell with acidic chloroform-methanol. Virology **54**, 369–383 (1973)

GRIMES, W. J., BURGE, B. W.: Modification of Sindbis virus glycoprotein by host-specified glycosyl transferases. J. Virol. **7**, 309–313 (1971)

GRIMLEY, P. M., FRIEDMAN, R. M.: Development of Semliki Forest virus in mouse brain: an electron microscopic study. Exp. molec. Path. **12**, 1–13 (1970)

HAGOPIAN, A., BOSMANN, H. B., EYLAR, E. H.: Glycoprotein biosynthesis: the localization of polypeptidyl: N-acetylgalactosaminyl, collagen:glucosyl, and glycoprotein:galactosyl transferases in HeLa cell membrane fractions. Arch. Biochem. Biophys. **128**, 387–396 (1968)

HARRISON, S. C., CASPAR, D. L. D., CAMERINI-OTERO, R. D., FRANKLIN, R. M.: Lipid and protein arrangement in bacteriophage PM2. Nature (Lond.) New Biol. **229**, 197–201 (1971a)

Harrison, S. C., David, A., Jumblatt, J., Darnell, J. E.: Lipid and protein organization in Sindbis virus. J. molec. Biol. 60, 523–528 (1971b)

Haslam, E. A., Hampson, A. W., Radiskevics, I., White, D. O.: The polypeptides of influenza virus. III. Identification of the hemagglutinin, neuraminidase and nucleocapsid proteins. Virology 42, 566–575 (1970)

Hayman, M. J., Skehel, J. J., Crumpton, M. J.: Purification of virus glycoproteins by affinity chromatography using *Lens culinaris* phytohemagglutinin. FEBS Letters 29, 185–188 (1973)

Heine, J. W., Roizman, B.: Proteins specified by Herpes simplex virus. IV. Contiguity of host and viral proteins in the plasma membrane of infected cells. J. Virol. 11, 810–813 (1973)

Hirano, H., Parkhouse, B., Nicolson, G. L., Lennox, E. S., Singer, S. J.: Distribution of saccharide residues on membrane fragments from a myeloma cell homogenate: its implications for membrane biogenesis. Proc. nat. Acad. Sci. (Wash.) 69, 2945–2949 (1972)

Holland, J. J., Kiehn, E. D.: Influenza virus effects on cell membrane proteins. Science 167, 202–205 (1970)

Homma, M., Tamagawa, S.: Restoration of the fusion activity of L cell-borne Sendai virus by trypsin. J. gen. Virol. 19, 423–426 (1973)

Hosaka, Y., Shimizu, Y. K.: Artificial assembly of envelope particles of HVJ (Sendai virus). I. Assembly of hemolytic and fusion factors from envelopes solubilized by Nonidet P40. Virology 49, 627–639 (1972a)

Hosaka, Y., Shimizu, Y. K.: Artificial assembly of envelope particles of HVJ (Sendai virus). II. Lipid components for formation of the active hemolysin. Virology 49, 640–646 (1972b)

Howatson, A. F.: Vesicular stomatitis and related viruses. Advanc. Virus Res. 16, 195–256 (1970)

Hoyle, L., Horne, R. W., Waterson, A. P.: The structure and composition of the myxoviruses. II. Components released from the influenza virus particle by ether. Virology 13, 448–459 (1961)

Hubbell, W. L., McConnell, H. M.: Spin label studies of the excitable membranes of nerve and muscle. Proc. nat. Acad. Sci. (Wash.) 61, 12–16 (1968)

Hubbell, W. L., Metcalfe, J. C., Metcalfe, S. M., McConnell, H. M.: The interaction of small molecules with spin-labelled erythrocyte membrane. Biochim. biophys. Acta (Amst.) 219, 415–427 (1970)

Israelachvili, J. N.: Theoretical considerations on the asymmetrical distribution of charged phospholipid molecules on the inner and outer layers of curved bilayer membranes. Biochim. biophys. Acta (Amst.) 323, 659–663 (1973)

Ji, T. H., Benson, A. A.: Association of lipids and proteins in chloroplast lamellar membrane. Biochim. biophys. Acta (Amst.) 150, 686–693 (1968)

Jost, P. C., Griffith, O. H., Capaldi, R. A., Vanderkooi, G.: Evidence for boundary lipid in membranes. Proc. nat. Acad. Sci. (Wash.) 70, 480–484 (1973)

Kaluza, G., Scholtissek, C., Rott, R.: Inhibition of the multiplication of enveloped RNA-viruses by glucosamine and 2-deoxy-D-glucose. J. gen. Virol. 14, 251–259 (1972)

Kang, C. Y., Prevec, L.: Proteins of vesicular stomatitis virus. II. Immunological comparisons of viral antigens. J. Virol. 6, 20–27 (1970)

Kates, M., Allison, A. C., Tyrrell, D. A. J., James, A. T.: Lipids of influenza virus and their relation to those of host cells. Biochim. biophys. Acta (Amst.) 52, 455–466 (1961)

Katz, E., Margalith, E.: Location of vaccinia virus structural polypeptides on the surface of the virus particle. J. gen. Virol. 18, 381–384 (1973)

Katz, E., Moss, B.: Formation of a vaccinia virus structural polypeptide from a higher molecular weight precursor: inhibition by rifampicin. Proc. nat. Acad. Sci. (Wash.) 66, 677–684 (1970)

Kauzmann, W.: Some factors in the interpretation of protein denaturation. Advanc. Protein Chem. 14, 1–63 (1959)

Kendal, A. P., Apostolov, K., Belyavin, G.: The effect of protease treatment on the morphology of influenza A, B and C viruses. J. gen. Virol. 5, 141–143 (1969)

Kingsbury, D. W.: Paramyxovirus replication. Curr. Topics Microbiol. and Immunol. 59, 1–34 (1972)

KLENK, H.-D.: Glycoproteins and glycolipids in viral envelopes. In: Membrane mediated information (P. W. KENT, ed.), vol. 1, p. 200–211. Lancaster: Medical and Technical Publishing Co. 1973

KLENK, H.-D., COMPANS, R. W., CHOPPIN, P. W.: An electron microscopic study of the presence or absence of neuraminic acid in enveloped viruses. Virology **42**, 1158–1162 (1970)

KLENK, H.-D., CHOPPIN, P. W.: Glycosphingolipids of plasma membranes of cultured cells and an enveloped virus (SV5) grown in these cells. Proc. nat. Acad. Sci. (Wash.) **66**, 57–64 (1970)

KLENK, H.-D., ROTT, R.: Formation of influenza virus proteins. J. Virol. **11**, 823–831 (1973)

KLENK, H.-D., ROTT, R., BECHT, H.: On the structure of the influenza virus envelope. Virology **47**, 579–591 (1972)

KORNBERG, R. D., McCONNELL, H. M.: Lateral diffusion of phospholipids in a vesicle membrane. Proc. nat. Acad. Sci. (Wash.) **68**, 2564–2568 (1971)

KUWERT, E., BÖHME, U., LICKFELD, K. G., BÖHME, W.: Zur Oberflächenstruktur des Tollwutvirion (TWV). Zbl. Bakt. I. Abt. Orig. A **219**, 39–45 (1972)

LAFFERTY, K. J.: The interaction between virus and antibody. II. Mechanism of the reaction. Virology **21**, 76–90 (1963)

LANDSBERGER, F. R., COMPANS, R. W., CHOPPIN, P. W., LENARD, J.: Organisation of the lipid phase in viral membranes. Effects of independent variation of the lipid and the protein composition. Biochemistry (Wash.) **12**, 4498–4502 (1973)

LANDSBERGER, F. R., LENARD, J., PAXTON, J., COMPANS, R. W.: Spin-label electron spin resonance study of the lipid-containing membrane of influenza virus. Proc. nat. Acad. Sci. (Wash.) **68**, 2579–2583 (1971)

LAVER, W. G.: Structural studies on the protein subunits from three strains of influenza virus. J. molec. Biol. **9**, 109–124 (1964)

LAVER, W. G.: Separation of two polypeptide chains from the hemagglutinin subunit of influenza virus. Virology **45**, 275–288 (1971)

LAVER, W. G.: The polypeptides of influenza viruses. Advanc. Virus Res. **18**, 57–103 (1973)

LAVER, W. G., BAKER, N.: Amino acid composition of polypeptides from influenza virus particles. J. gen. Virol. **17**, 61–67 (1972)

LAVER, W. G., VALENTINE, R. C.: Morphology of the isolated hemagglutinin and neuraminidase subunits of influenza virus. Virology **38**, 105–119 (1969)

LAZAROWITZ, S. G., COMPANS, R. W., CHOPPIN, P. W.: Influenza virus structural and non-structural proteins in infected cells and their plasma membranes. Virology **46**, 830–843 (1971)

LAZAROWITZ, S. G., GOLDBERG, A. R., CHOPPIN, P. W.: Proteolytic cleavage by plasmin of the HA polypeptide of influenza virus: host cell activation of serum plasminogen. Virology **56**, 172–180 (1973)

LENARD, J., COMPANS, R. W.: The membrane structure of lipid-containing viruses. Biochim. biophys. Acta (Amst.) **344**, 51–94 (1974)

LENARD, J., WONG, C. Y., COMPANS, R. W.: Association of internal membrane protein with the lipid bilayer in influenza virus. Biochim. biophys. Acta (Amst.) **332**, 341–349 (1974)

LESSLAUER, W., CAIN, J. E., BLASIE, J. K.: X-ray diffraction studies of lecithin bimolecular lipid leaflets with incorporated fluorescent probes. Proc. nat. Acad. Sci. (Wash.) **69**, 1499–1503 (1972)

LEVITAN, D. B., BLOUGH, H. A.: Preliminary biochemical characterization of the herpesvirus fusion factor(s). Virology (in press)

McSHARRY, J. J., COMPANS, R. W., CHOPPIN, P. W.: Proteins of vesicular stomatitis virus and of phenotypically mixed vesicular stomatitis virus and simian virus 5 virions. J. Virol. **8**, 722–729 (1971)

McSHARRY, J. J., COMPANS, R. W., LACKLAND, H., CHOPPIN, P. W.: Isolation of viral membrane proteins, p. 215. Abstr. 72nd Ann. Mtg. Amer. Soc. Microb. Phila., 1972

McSHARRY, J. J., WAGNER, R. R.: Lipid composition of purified vesicular stomatitis viruses. J. Virol. **7**, 59–70 (1971)

Melchers, F.: Biosynthesis, intracellular transport, and secretion of immunoglobulins. Effect of 2-deoxy-D-glucose in tumor plasma cells producing and secreting immunoglobulin G_1. Biochemistry (Wash.) **12**, 1471–1476 (1973)

Michaelson, D. M., Horwitz, A. F., Klein, M.: Transbilayer asymmetry and surface homogeneity of mixed phospholipids in cosonicated vesicles. Biochemistry (Wash.) **12**, 2637–2645 (1973)

Mora, P. T., Cumar, F. A., Brady, R. O.: A common biochemical change in SV40 and polyoma virus transformed mouse cells coupled to control of cell growth in culture. Virology **46**, 60–72 (1971)

Mudd, J. A., Summers, D. F.: Protein synthesis in vesicular stomatitis virus infected HeLa cells. Virology **42**, 328–340 (1970)

Nermut, M. V.: Further investigation on the fine structure of influenza virus. J. gen. Virol. **17**, 317–331 (1972)

Nermut, M. V., Frank, H.: Fine structure of influenza A_2 (Singapore) as revealed by negative staining, freeze-drying and freeze-etching. J. gen. Virol. **10**, 37–51 (1971)

Nermut, M. V., Frank, H., Schäfer, W.: Properties of mouse leukemia viruses. III. Electron microscopic appearance as revealed after conventional preparation techniques as well as freeze-drying and freeze-etching. Virology **49**, 345–358 (1972)

Neurath, A. R., Vernon, S. K., Dobkin, M. B., Rubin, B. A.: Characteristics of subviral components resulting from treatment of rabies virus with tri(n-butyl) phosphate. J. gen. Virol. **14**, 33–48 (1972)

Oda, K., Joklik, W. K.: Hybridization and sedimentation studies on "early" and "late" vaccinia messenger RNA. J. molec. Biol. **27**, 395–419 (1967)

Ohki, S., Aono, O.: Phospholipid bilayer-micelle transformation. J. Colloid and Interface Sci. **32**, 270–281 (1970)

Okada, Y., Kim, J.: Interaction of concanavalin A with enveloped viruses and host cells. Virology **50**, 507–515 (1972)

Pfefferkorn, E. R., Hunter, H. S.: The source of the ribonucleic acid and phospholipid of Sindbis virus. Virology **20**, 446–456 (1963)

Phillips, D. R., Morrison, M.: Exposed protein on the intact human erythrocyte. Biochemistry (Wash.) **10**, 1766–1771 (1971)

Pinto da Silva, P., Branton, D.: Membrane splitting in freeze-etching. J. Cell Biol. **45**, 598–605 (1970)

Printz, P., Wagner, R. R.: Temperature-sensitive mutants of vesicular stomatitis virus: synthesis of virus-specific proteins. J. Virol. **7**, 651–662 (1971)

Ray, E. K., Blough, H. A.: Biosynthesis and transport of herpesvirus glycoproteins and glycolipids., p. 202 Abstr. 74th Ann. Mtg. Amer. Soc. Microbiol. Chicago, 1974

Reginster, M., Calberg-Bacq, C.-M.: Further observations on the effects of caseinase C on the envelope of influenza and Newcastle disease viruses. J. Ultrastruct. Res. **23**, 144–152 (1968)

Reimer, C. B., Baker, R. S., Newlin, T. E., Havens, M. L.: Influenza virus purification with the zonal ultracentrifuge. Science **152**, 1379–1381 (1966)

Rudy, B., Gitler, C.: Microviscosity of the cell membrane. Biochim. biophys. Acta (Amst.) **288**, 231–236 (1972)

Salem, L.: Attractive forces between macromolecular chains of biological importance. Nature (Lond.) **193**, 476–477 (1962a)

Salem, L.: The role of long-range forces in the cohesion of lipoproteins. Canad. J. Biochem. **40**, 1287–1298 (1962b)

Sarov, I., Joklik, W. K.: Studies on the nature and location of the capsid polypeptides of vaccinia virions. Virology **50**, 579–592 (1972)

Scanu, A. M., Tardieu, A.: Temperature transitions of lipid mixtures containing cholesterol esters. Relevance to the structural problem of serum high density lipoprotein. Biochim. biophys. Acta (Amst.) **231**, 170–174 (1971)

Schaefer, R., Hinnen, R., Franklin, R. M.: Further observations on the structure of the lipid-containing bacteriophage PM2. Nature (Lond.) **248**, 681–682 (1974)

Scheid, A., Choppin, P. W.: Identification of biological activities of paramyxovirus glycoproteins. Activation of cell fusion, hemolysis, and infectivity by proteolytic cleavage of an inactive precursor protein of Sendai virus. Virology **57**, 475–490 (1974)

SCHLESINGER, M. J., SCHLESINGER, S., BURGE, B. W.: Identification of a second glycoprotein in Sindbis virus. Virology **47**, 539–541 (1970)
SCHULZE, I. T.: The structure of influenza virus. I. The polypeptides of the virion. Virology **42**, 890–904 (1970)
SCHULZE, I. T.: The structure of influenza virus. II. A model based on the morphology and composition of subviral proteins. Virology **47**, 181–196 (1972)
SCHULZE, I. T.: Structure of the influenza virion. Advanc. Virus Res. **18**, 1–56 (1973)
SEBRING, E. D., SALZMAN, N. P.: Metabolic properties of early and late vaccinia virus messenger ribonucleic acid. J. Virol. **1**, 550–558 (1967)
SEGREST, J. P., JACKSON, R. L., ANDREWS, E. P., MARCHESI, V. T.: Human erythrocyte membrane glycoprotein: a re-evaluation of the molecular weight as determined by SDS polyacrylamide gel electrophoresis. Biochem. biophys. Res. Commun. **44**, 390–395 (1971)
SETO, J. T., CHANG, F. S.: Functional significance of sialidase during influenza virus multiplication: an electron microscope study. J. Virol. **4**, 58–66 (1969)
SIMONS, K., KERÄNEN, S., KÄÄRIÄINEN, L.: Identification of a precursor for one of the Semliki Forest virus membrane proteins. FEBS Letters **29**, 87–91 (1973)
SINGER, S. J., NICOLSON, G. L.: The fluid mosaic model of the structure of cell membranes. Science **175**, 720–731 (1972)
SKEHEL, J. J.: Estimations of the molecular weight of the influenza virus genome. J. gen. Virol. **11**, 103–109 (1971)
SKEHEL, J. J., SCHILD, G. C.: The polypeptide composition of influenza A viruses. Virology **44**, 396–408 (1971)
SOKOL, F., CLARK, H. F.: Phosphoproteins, structural components of rhabdoviruses. Virology **52**, 246–263 (1973)
SOKOL, F., STANCEK, D., KOPROWSKI, H.: Structural proteins of rabies virus. J. Virol. **7**, 241–249 (1971)
SPEAR, P. G., ROIZMAN, B.: Proteins specified by Herpes simplex virus. V. Purification and structural proteins of the virus. J. Virol. **9**, 143–159 (1972)
STANLEY, P., CROOK, N. E., STREADER, L. G., DAVIDSON, B. E.: The polypeptides of influenza virus. VIII. Large-scale purification of the hemagglutinin. Virology **56**, 640–645 (1973)
STANLEY, P., HASLAM, E. A.: The polypeptides of influenza virus. V. Localization of polypeptides in the virion by iodination techniques. Virology **46**, 764–773 (1971)
STRAND, M., AUGUST, J. T.: Protein kinase and phosphate acceptor proteins in Rauscher murine leukamia virus. Nature (Lond.) New Biol. **233**, 137–140 (1971)
THOMPSON, T. E., SEARS, B.: Non-ideal mixing of phosphatidylcholine and cholesterol in single-walled bilayer vesicles. (Abstract.) Fed. Proc. **33**, 1551 (1974)
TIFFANY, J. M., BLOUGH, H. A.: Myxovirus envelope proteins: a directing influence on the fatty acids of membrane lipids. Science **163**, 573–574 (1969a)
TIFFANY, J. M., BLOUGH, H. A.: Fatty acid composition of three strains of Newcastle disease virus. Virology **37**, 492–494 (1969b)
TIFFANY, J. M., BLOUGH, H. A.: Models of structure of the envelope of influenza virus. Proc. nat. Acad. Sci. (Wash.) **65**, 1015–1112 (1970a)
TIFFANY, J. M., BLOUGH, H. A.: Estimation of the number of surface projections of myxo- and paramyxoviruses. Virology **41**, 392–394 (1970b)
TREMAINE, J. H., GOLDSACK, D. E.: The structure of regular viruses in relation to their subunit amino acid composition. Virology **35**, 227–237 (1968)
VERKLEIJ, A. J., ZWAAL, R. F. A., ROELOFSEN, B., COMFURIUS, P., KASTELIJH, D., VAN DEENEN, L. L. M.: The asymmetric distribution of phospholipids in the human red cell membrane. A combined study using phospholipases and freeze-etch electron microscopy. Biochim. biophys. Acta (Amst.) **323**, 178–193 (1973)
VERNON, S. K., NEURATH, A. R., RUBIN, B. A.: Electron microscopic studies on the structure of rabies virus. J. Ultrastruct. Res. **41**, 29–42 (1972)
WAGNER, R. R., PREVEC, L., BROWN, F., SUMMERS, D. F., SOKOL, F., MACLEOD, R.: Classification of rhabdovirus proteins: a proposal. J. Virol. **10**, 1228–1230 (1972)
WAITE, M. R., BROWN, D. T., PFEFFERKORN, E. R.: Inhibition of Sindbis virus release by media of low ionic strength: an electron microscope study. J. Virol. **10**, 537–544 (1972)

Weinstein, D. B., Blough, H. A.: Kinetics of incorporation of ^{32}P into several cell lines and the inability to turn it over. Biochim. biophys. Acta (Amst.) (in press)

Wildy, P.: Classification and nomenclature of viruses. Monogr. Virol. 5, 1–81 (1971)

Wilkins, M. H. F., Blaurock, A. E., Engelman, D. M.: Bilayer structure in membranes. Nature (Lond.) New Biol. 230, 72–76 (1971)

Wirtz, K. W., Zilversmit, D. B.: Exchange of phospholipids between liver mitochondria and microsomes *in vitro*. J. biol. Chem. 243, 3596–3602 (1968)

Wu, C.-W., Stryer, L.: Proximity relationships in rhodopsin. Proc. nat. Acad. Sci. (Wash.) 69, 1104–1108 (1972)

Zingsheim, H. P.: Membrane structure and electron microscopy. The significance of physical problems and techniques (freeze-etching). Biochim. biophys. Acta (Amst.) 265, 339–366 (1972)

Zwaal, R. F. A., Roelofsen, B., Comfurius, P., van Deenen, L. L. M.: Complete purification and some properties of phospholipase C from *Bacillus cereus*. Biochim. biophys. Acta (Amst.) 233, 474–479 (1971)

Latent Herpes Simplex Virus and the Nervous System

Jack G. Stevens[1]

With 5 Figures

Table of Contents

I. Introduction

The natural history of herpetic disease is a most intriguing phenomenon in virology, and its definition has consumed the activities of a diverse group of individuals for the past three quarters of a century. The early suspicion that the nervous system plays a crucial role is now well-documented and appreciated. The work which has led to our current knowledge of this association is the subject of this review.

II. Historical

Suggestions that a relationship exists between cutaneous herpetic lesions and the nervous system (more specifically the trigeminal ganglion) first appeared at the turn of the 19th century. At that time, Howard was sufficiently impressed by the classic work of Head and Campbell [which indicated that zoster was somehow related to sensory ganglia (1900)] and his own

---

1 Reed Neurological Research Center and Department of Microbiology and Immunology. School of Medicine, University of California, Los Angeles, California 90024.

observations on coincident herpetic lesions, trigeminal ganglionitis, and pneumonitis to publish 2 papers of documentation (1903, 1905). Also in 1905, in
the course of an extensive presentation involving twenty cases of trigeminal
neuralgia which had been treated by extirpation of the corresponding ganglion,
CUSHING noted that two individuals developed herpetic lesions in areas
supplied by the contralateral nerve, but no lesions on the ipsilateral side. This
observation suggested that an intact trigeminal tract was necessary for the
development of herpetic lesions. These limited observations by HOWARD and
himself led CUSHING to the assertion that "posterior-root ganglia lesions are
responsible for the common forms of herpes about the eyes and nose" (CUSHING,
1904). It is important to note that these authors dealt with herpetic disease
purely as a clinical and pathologic entity, with no reference or discussion
being given to the etiology. This had in fact been studied in a preliminary
way by VIDAL (1873) who, by re-inoculation of lesion-material into the same
person, established the infectious nature of the disease.

Although CUSHING'S conclusions were provocative, little further interest
was shown in either the infectious agent or the various manifestations of the
disease until the end of World War I when GRÜTER (1920) and LÖWENSTEIN
(1919) succeeded in producing herpetic keratitis in rabbits. This was a most
important finding, for it established the rabbit as a useful experimental
animal, led immediately to the observations that nervous symptoms sometimes developed in rabbits with herpetic keratitis (DOERR, 1920; DOERR and
VÖCHTING, 1920), and prompted the extensive and insightful experiments
described over the next 10 year period by GOODPASTURE and TEAGUE. These
experiments were among the most important in establishing present day
concepts concerning the pathogenesis of disease induced by Herpes simplex
virus (HSV), and the general correctness of the conclusions and extensions
is even more remarkable when one considers the experimental techniques
available at the time.

Earlier workers had suggested that corneal inoculation led to the central
nervous system (CNS) invasion through either the bloodstream (DOERR and
SCHNABEL, 1921) or the optic nerve (LEVADITI and HARVIER, 1920). The observation by DOERR and VÖCHTING (1920) that rabbits which developed encephalitis
after unilateral corneal inoculation of the virus constantly twisted their heads to
the affected side suggested to GOODPASTURE and TEAGUE that viremia was not
the route by which the brain was infected, and experiments were therefore
designed to carefully study the pathogenesis of encephalitis following corneal
inoculation. It was shown (GOODPASTURE and TEAGUE, 1923), that animals
inoculated by this route subsequently developed grossly apparent lesions in the
trigeminal tract of the pons and medulla on the ipsilateral side, but no lesions
were observed in the optic tract. Microscopically, polymorphonuclear and mononuclear infiltrations were noted in the affected area, and characteristic intranuclear inclusion bodies were seen in the glia and neurons. In addition, similar
results obtained when the conjuctiva were inoculated following enucleation
of the eye. Although no specific lesions were observed in either the trigeminal

nerve or ganglion, GOODPASTURE and TEAGUE concluded that the infection most probably reached the central nervous system through the trigeminal nerve[2].

As additional evidence to support the concept that HSV travels along nerve trunks, GOODPASTURE inoculated rabbits in several sites and was able to demonstrate CNS lesions in areas corresponding to the central terminations of not only other sensory nerves, but also motor and sympathetic nerves. For example, inoculation of the masseter muscle induced an encephalitis which was localized in the ipsilateral portion of the pons and medulla, with destruction of the motor nucleus of the corresponding trigeminal nerve. Inoculation of the ovary was followed by posterior paralysis and acute mid-thoracic myelitis. Although intramuscular injection into a hind leg normally produced paralysis, similar injections made distal to a severed sciatic nerve resulted in no disease. With the completion of these experiments, the relationship between peripheral infection and central nervous system disease was fully established, and a strong argument for transmission in nerve trunks was made. GOODPASTURE further suggested that the most likely intraneural route taken was an axonal one (GOODPASTURE, 1925 a). As will be seen later, these conclusions have survived the more rigorous investigations performed in subsequent years. Finally, in addition to the experiments described, GOOD-PASTURE presented the first evidence concerning natural route of infection and the possible serious consequences of such infection. Here, GOODPASTURE (1925 b) observed that rabbits kept in contact with other animals presenting herpetic keratitis developed a herpetic encephalitis which was first associated with the central nuclei of the trigeminal and glossopharyngeal nerves. The inference was made that the portal of entry was mucosa of the mouth, nose, and throat.

Even though the nature of viruses had not yet been established, by this time it was clear that herpetic disease was induced by a transmissible agent with the properties ascribed at that time to viruses, and GOODPASTURE concluded (1929) that "it seems to me probable from experimental and clinical facts that herpetic virus does reside in a latent state within the human body and specifically in the nervous tissues, perhaps primarily within nerve cells of the ganglion, and that neural disturbances are frequently the basis of subsequent outbreaks".

The next year, ANDREWES and CARMICHAEL (1930) showed unequivocally that some 3/4 of individuals sampled randomly from the population had significant levels of neutralizing antibody to herpes simplex virus. In addition, and most importantly, those individuals who suffered from recurrent herpetic

---

2 Although I have chosen to discuss the work of GOODPASTURE and TEAGUE in some detail, it is important to note that in the same year MARINESCO and DRAGENESCO (1923) presented the results of experiments concerning the pathogenesis of herpetic disease in rabbits. In this completely independent investigation, they also concluded that the virus traveled in nerve trunks. The work of GOODPASTURE and TEAGUE was chosen because it is more extensive, and was followed by additional significant findings and important conclusions by the same authors.

lesions also were shown to possess specific neutralizing antibody. This finding
was a peculiar one, for it was not at all consistent with the established concepts
of infectious diseases. This difficult conceptual problem partially disappeared
when Dodd et al. (1938) and Burnet and Williams (1939) demonstrated
conclusively that the primary infection with herpes simplex often presented
an aphthous stomatitis, and that this was followed by appearance of neutralizing antibody. Similar proofs for primary infections in other areas of the
body have been reviewed and documented by Scott and Tokumaru (1965).
Thus, the disease was established as a typical infectious malady in which the
initial infection is followed by a specific immune response. However, the basis
for recurrences was obviously still not clear and, although they clearly favored
a mechanism based upon reactivation of the virus by various internal
and external stimuli, Burnet and Williams were noncommittal as to
whether the virus persisted in the nervous system, skin, salivary glands, or
mouth.

One additional set of observations is of considerable importance to the
historical development. The results of trigeminal neurectomy and various
other manipulations of the trigeminal tract as they relate to herpetic disease
were considered by Carton and Kilbourne (1952) and later by Carton
(1953). The following relevant conclusions were presented by Carton:
1) Following transection of the proximal trigeminal nerve root for trigeminal
neuralgia, herpetic lesions appear within the 48th to 96th hour after operation
somewhere in the peripheral area anesthetized by the operation in essentially
all patients operated. 2) These lesions do not appear if the nerve or its branches
are sectioned distal to the ganglion. 3) If interruption of the second and third
peripheral divisions of the nerve are followed by sectioning of the root,
cutaneous or mucosal lesions do not appear. 4) Injection of alcohol into the
ganglion (which results in destruction of neurons) is usually *not* followed by
appearance of herpetic lesions. Paine (1964) summarized and concluded from
these observations that "disturbing the posterior sensory root of the fifth
cranial nerve by operation results in the appearance of herpetic vesicles in
the areas innervated by the second and third peripheral divisions of the nerve
in a highly significant number of patients, provided the ganglion has not
been destroyed and the peripheral divisions are intact". Although Carton
and Kilbourne suggested that these manipulations resulted in activation of
virus from the skin, Paine indicated that the ganglion was a more likely site
in which the latent virus is maintained.

Taken together, the observations and experiments presented to this point
lead to the following general hypothesis which explains the natural history
of recurrent herpetic infection: Primary infection would result in viral
replication in epithelial cells of the skin or mucous membrane, and subsequent invasion of superficial nerve endings. The virus would then travel
intra-axonally in sensory nerves to the corresponding sensory ganglion (most
often the trigeminal) where a latent infection would be established, most
probably in neurons. Upon "activation" the virus would travel centrifugally

from the neuronal soma in axons, ultimately reaching the epithelium where lesions would again be produced[3].

The remainder of this review will be concerned with the more recent experimental and clinical evidence which supports this concept. In these considerations, results derived from experimental systems will be discussed interchangeably with observations made in man. At this juncture, such generalizations seem justified, since, where information is available, the systems are analagous.

III. Evidence That Infection of the Externa is Followed by Latent Infection of Sensory Ganglia

That an infection of skin, cornea or mucous membranes is followed by a latent infection of the corresponding sensory ganglia is now well established. The first and most extensively studied experimental system is a murine model which has served as the primary object of study in our own laboratories (STEVENS and COOK, 1971, 1973 a, b). In this system, infection of a rear footpad with HSV types 1 or 2 is followed by an acute infection of sciatic nerve, sacrosciatic spinal ganglia, dorsal roots, spinal cord, and brain. During the acute disease, infectious virus can be easily recovered from all these tissues. Here, it is of particular importance to note that in the spinal ganglia (Fig. 1) infectious virus first appears by the second day, reaches a peak by day 4, and disappears by day 7.

All mice that survive this initial infection and demonstrate a posterior paralysis (from which most undergo a clinical recovery) harbor latent virus

---

3 As work has progressed on recurrent herpetic disease, at least 2 alternate hypotheses have at times been seriously considered. First, it has been proposed that the virus is not latent, but that each recurrence results from reinfection, either with exogenous virus, or from virus which is continuously being replicated and released at appropriate sites. In the latter instance, lacrimal or salivary glands have been suggested to be the site of chronic infection (KAUFMAN et al., 1967). Although it is clear that certain individuals do intermittently shed the virus, even in the absence of lesions (cf DOUGLAS and COUCH, 1970) if this is the source of virus for recurrent lesions, it is difficult to understand how a continuously released endogenous virus picks only certain times to cause lesions, why such lesions usually reappear in the same site, and why individuals can often predict when they will appear (SCOTT and TOKUMARU, 1965; SCOTT, 1957). The last 2 arguments could also be applied to an exogenous infection.

The second hypothesis states that the virus is latent, but that it is latent in the same superficial tissues which demonstrate lesions. A limited number of transplantation experiments (NICOLAU and POINCLOUX, 1928; STALDER and ZURUKZOGLU, 1936) tend to rule against this interpretation. Of greater significance, however, are several extensive attempts (in the interval between recurrences) to detect latent or active virus in the tissues at the site of recurrent lesions (FINDLAY and MACCALLUM, 1940; CORIELL, 1963; RUSTIGIAN et al., 1966). In none of these experiments could virus be demonstrated when direct isolation was attempted or when the tissues were maintained as organ cultures *in vitro*. It is noteworthy that the procedure of organ culture is effective in "inducing" virus from latently infected sensory ganglia (see next section).

                                J. G. Stevens:

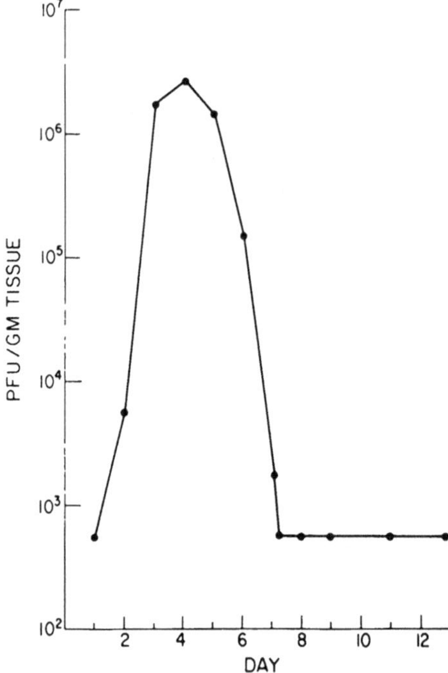

Fig. 1. Replication cycle of Herpes simplex virus in sciatic spinal ganglia of mice previously inoculated in a rear footpad. Four mice were sacrificed at each time period, spinal ganglia were dissected from the animal, pooled and assayed directly for infectious virus. The level of sensitivity for the assay is 500 PFU/gm of tissue. From STEVENS and COOK (1973a)

in the ipsilateral sacrosciatic spinal ganglia, many in the contralateral ganglia, and some in the spinal cord. In contradistinction to the ganglia, virus does not appear to be harbored in the associated nerve trunks. These latent infections probably persist for the life of the mouse. The secret to demonstrating latent virus lies in explantation and cultivation of these tissues as organ cultures *in vitro*. When this is done, within a few days infectious virus appears in the explants (Fig. 2) and culture media. The "reactivated" virus can then be detected either by subsequent inoculation of indicator monolayer cell cultures, or inclusion of the viral sensitive cells in the same vessel in which the tissues are incubated. At the time of explant, neither infectious virus, viral specific antigens, nor ultrastructural evidence of viral specific products can be demonstrated in these tissues. Recently (STEVENS and COOK, 1974) by employing *in situ* nucleic acid hybridization techniques, we have been able to detect viral DNA in neuronal somas populating such ganglia, and this will be discussed in greater detail later.

After latent virus had been found in murine spinal ganglia, by employing similar methods of organ cultivation, we succeeded in demonstrating latent virus in the trigeminal ganglia (STEVENS et al., 1972) and brainstems (KNOTTS

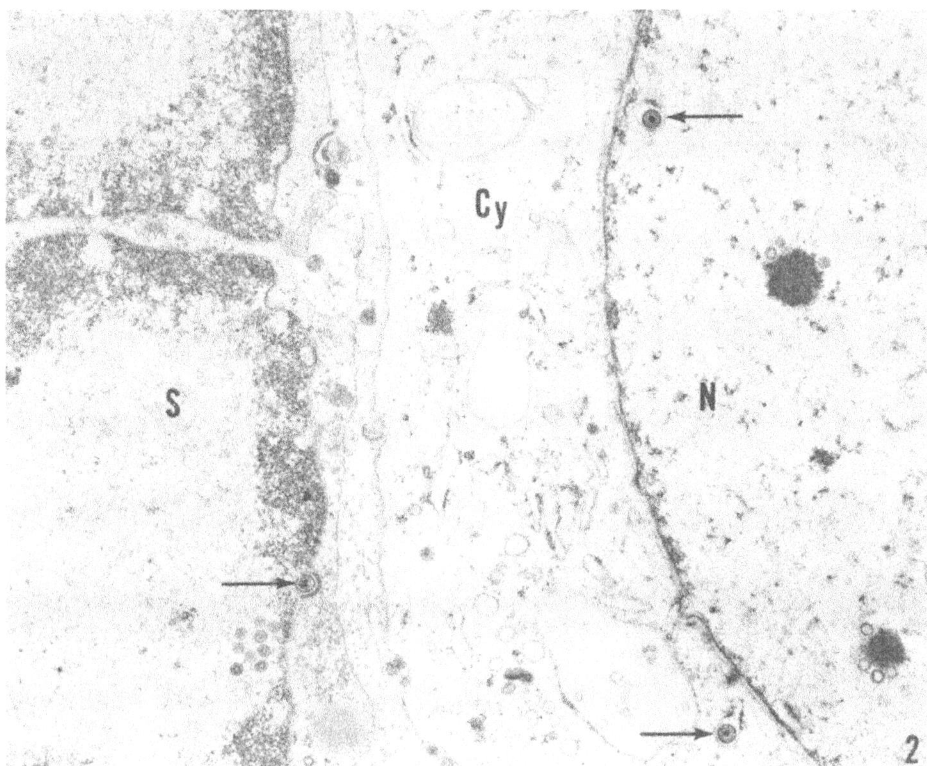

Fig. 2. Two virus-infected cells from a spinal ganglion that had been in culture 5 days prior to processing for electron microscopy. On the right, the nucleus (N) of a neuron contains a virion (arrow) and two dense bodies partly surrounded by viral capsids. Another virion (arrow) is seen in the cytoplasm (Cy) of this cell. The nucleus of the supporting cell (S) on the left encloses a virion (arrow) as well as viral capsids. ×16800. From STEVENS and COOK (1973a)

et al., 1973) of rabbits previously inoculated on scarified corneas. These techniques were applied by others, and in rapid succession, the findings were extended by at least 3 laboratories to natural infections involving the trigeminal ganglion in man (BASTION et al., 1972; BARINGER and SWOVELAND, 1973; RADDA et al., 1973). Here, particularly from the more extensive data presented by BARINGER and SWOVELAND (1973) and BARINGER (1974, and personal communication) it now appears likely that $^1/_2$ or more of the human population harbors latent Herpes simplex virus in the trigeminal ganglion. More recently, Herpes simplex virus has been recovered from human sacral spinal ganglia (BARINGER, 1974), and WALZ et al. (1974) have succeeded in establishing latent infections of sensory ganglia in mice inoculated by various external routes. Of particular significance in the latter investigation was the success in detecting latent virus in lumbosacral ganglia of mice previously inoculated intravaginally with HSV-2. This finding has obvious significance concerning the pathogenesis of recurrent genital herpes lesions, and could

Table 1. Recovery of infectious herpes simplex virus after *in vitro* cultivation of latently infected brain stem from rabbits

Rabbit number	Months post infection	HSV latent in trigeminal ganglia	HSV latent in brain stem
1	3	+	+
2	3	+	−
3	3	+	−
4	1	+	−
5	1	+	−
6	1	+	+
7	1	+	+
8	2	−	−
9	2	−	−
10	12	−	−
11	5	+	−
12	12	+	+
13	9	+	−
14	9	+	+
Totals		11/14	5/14

Rabbits were inoculated on one scarified cornea with HSV. Animals were then killed at intervals, and tissues were removed, divided, and cocultivated with RK_{13} cells. Viral induced cytopathic effects in the RK_{13} cells were scored at daily intervals and isolates were identified immunologically as HSV. From Knotts et al. (1973).

take on great importance if Herpes simplex virus is ultimately shown to be etiologically related to cervical carcinoma.

The simple technique of organ culture has thus resolved the central problem attendant to relating sensory ganglia, latent Herpes simplex virus, and recurrent herpetic disease. Following infection of the skin, mucous membranes, or cornea, Herpes simplex viruses travel to, and are harbored in sensory ganglia.

IV. Evidence That Herpes Simplex Viruses Travel in Nerves

As summarized previously, the early work of Goodpasture suggested strongly that Herpes simplex virus travels in nerve trunks. Additional evidence was derived from two laboratories a decade ago. In one series of experiments, Johnson (1964) showed that footpad inoculation of mice was followed by an advancing, centripetal infection of endoneural cells in the sciatic nerve. In the other, Peter Wildy (1967) assayed infectious virus within segments of the peripheral nervous system after similar inoculations and showed that the infection traveled centripetally to and through the CNS. If the sciatic nerve was sectioned before inoculation, such passage did not occur. More recently, we expanded upon Wildy's experiments by titrating additional segments of the nerve and have shown a centripetal progression of virus which goes from foot to sciatic nerve to ganglia, to dorsal root, to CNS (Fig. 3). Evidence supporting the concept that virus moves centrifugally as well as centripetally in nerves will be presented in a later section.

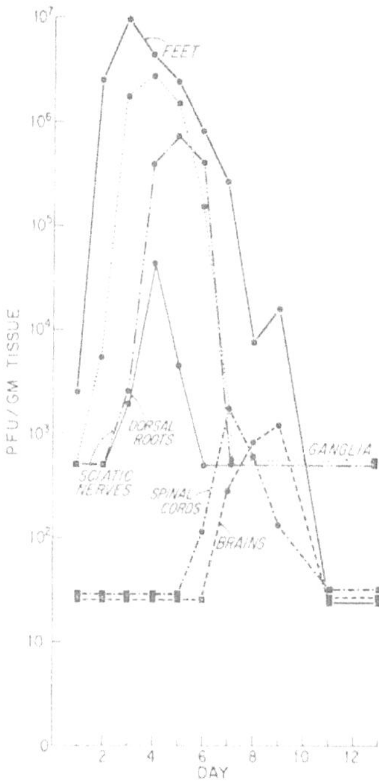

Fig. 3. Replication cycle of Herpes simplex virus in various tissues of mice previously inoculated in a rear footpad. Four mice were sacrificed at each time period, the various tissues were dissected from the animals, like tissues were pooled and assayed directly for infectious virus. The squares on all curves represent levels of sensitivity for the assay. Virus could not be recovered at any time from draining lymph nodes (sensitivity < 50 PFU/gm), blood serum (sensitivity < 10 PFU/ml), or ventral roots (sensitivity < 500 PFU/gm). In addition, neutralizing antibody became detectible in sera by the sixth day after infection. From Cook and Stevens (1973)

These experiments constitute strong support for the concept that centripetal passage occurs in nerves, and this concept is now generally accepted. However, the endoneural route taken is somewhat less clear. Goodpasture originally suggested that replicating virus moved intra-axonally. Although his conclusion concerning movement by replication is obviously untenable (HSV needs a nucleus for replication), the bulk of present evidence indicates that the virus does travel centripetally in axons. As is detailed elsewhere (Cook and Stevens, 1973), the experimental evidence supporting this conclusion derives from several sources, but relates principally to a) the fact that axonal flow is now known to be bi-directional (Zelená, 1969; Lavail and Lavail, 1972), b) to the speed (≥ 1.5 mm/hour) with which the infection moves in nerves (even in the presence of large amounts of anti-viral antibody), and c) to the finding by Kristensson et al. (1971) that the infection progresses even in nerves of mice treated with colchicine or vinblastine (substances which disrupt neurotubules). It is noteworthy that the speed with which the

virus travels to the ganglia is not dissimilar to that ($\sim$3 mm/hr) derived by
Lavail and Lavail (1972) for the speed of reverse axonal flow.

V. Characteristics of the Latent Infection in Sensory Ganglia
A. Cell-type Involved

During the acute infection, it has been shown by several laboratories
(Cook and Stevens, 1973; Dillard et al., 1972; Schwartz and Elizan,
1973; Knotts et al., 1974) that Schwann, satellite, and other supporting
cells in experimental animals undergo an abortive infection, and that morpho-
logically complete virions are replicated in neurons. From these observations,
it might be suggested that the supporting cells would be the most likely
cell-type to harbor latent virus. However, this appears not to be the case;
the bulk of evidence indicates that latent virus is associated with neurons.
To establish unequivocally which cells do harbor virus, the various cell-types
in the ganglia must be separated, incubated, and the cultures which produce
virus assessed. In our hands, such a separation has not been achieved, and
we have therefore relied on several indirect experiments to indicate the cell-
type involved in harboring virus. All these experiments are consistent with a
neuronal site. The work has been recently published (Cook et al., 1974), and
can be summarized as follows: 1) As stated previously, ganglia selectively
harbor virus; sciatic nerve trunks do not harbor the agent even though virus
passes through them during the acute infection and they are capable of
replicating virus when infected *in vitro*. Since the neuronal soma is the only
cell-type present in ganglia which is not also present in nerve trunks, the
soma is implicated as the reservoir of virus. 2) Latently infected ganglia
transplanted to noninfected syngeneic mice are induced to make infectious
virus. When these ganglia were subjected to immunofluorescent methods
specific for viral antigens, it was found that the initial cells producing viral
antigens were isolated neurons. The infection then spread to adjacent supporting cells. 3) When the immunofluorescent methods were replaced by ultrastructural techniques, the neuron was identified as the first cell in which
viral specific products could be detected. 4) Most convincingly, experiments
were performed which identified the neuron as the first cell in which replicating
viral DNA could be detected. Initially, explanted, latently infected ganglia
were maintained *in vitro* in the presence of ^{3}H-thymidine. At appropriate
intervals, frozen sections were made and autoradiographs were prepared.
When these preparations were examined microscopically, it was found that
concentrations of silver grains were selectively present over isolated neurons
(Fig. 4a). In addition, when these ^{3}H-thymidine "pulses" were followed by
"chases" with a 100-fold excess of unlabeled thymidine, the grains appeared
in supporting cells surrounded by degenerated neurons (Fig. 4b). Finally, the
DNA in the "reactivating" neurons was identified as viral DNA by *in situ*
nucleic acid hybridization experiments (Fig. 4c).

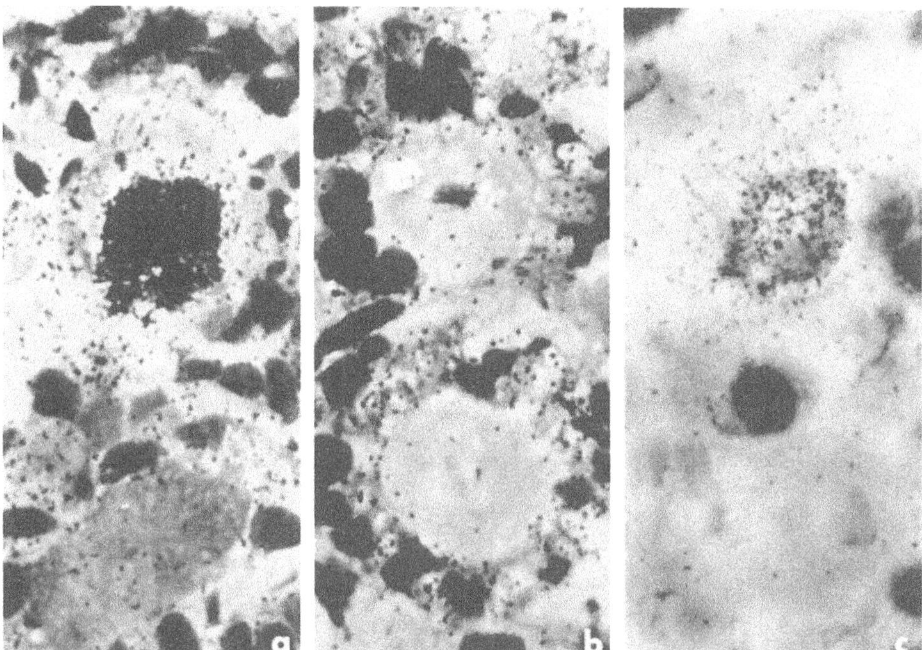

Fig. 4a–c. Autoradiographs of sectioned spinal ganglia processed after "reactivation" of latent herpes simplex virus *in vitro*. a A spinal ganglion maintained *in vitro* for 24 hours in the presence of ³H-thymidine. The heavy concentration of silver grains over the nucleus of the upper neuron can be compared with a lower neuron which exhibits background labeling. ×900. b A spinal ganglion which was maintained *in vitro*, "pulsed" with ³H-thymidine for 24 hours, and "chased" with excess unlabeled thymidine for an additional 12 hours. Silver grains can be seen at the periphery of two neurons and in the satellite cells surrounding them. ×900. c A latently infected spinal ganglion processed after 63 hours of incubation *in vitro*. After sectioning, the specimen was subjected to *in situ* nucleic acid hybricization procedures involving radioactive Herpes simplex virus specific complementary RNA. A concentration of silver grains denoting the presence of Herpes simplex virus DNA is seen over a neuron. This can be compared with the unlabeled neuron below. ×900. From Cook et al. (1974)

B. State of the Virus Associated with Neurons

As has been described in some detail by Roizman (1965, 1974), the persistence of Herpes simplex virus can be explained by two alternate hypotheses. In the first (the dynamic state hypothesis), the infection would persist as infectious virus. Here, a small number of cells would constantly replicate virus, and the infection would be localized by antibody, immune lymphocytes, or possibly, interferon. The alternative explanation (static state hypothesis) involves conservation of the viral genome in some non-replicating state. Here, the viral DNA could be maintained either in some extrachromosomal site or integrated into the cellular DNA. The data collected to date do not permit an absolute discrimination between dynamic and static state hypotheses but results from several types of experiment are most consistent with the latter. First, it was shown that no infectious virus could be recovered

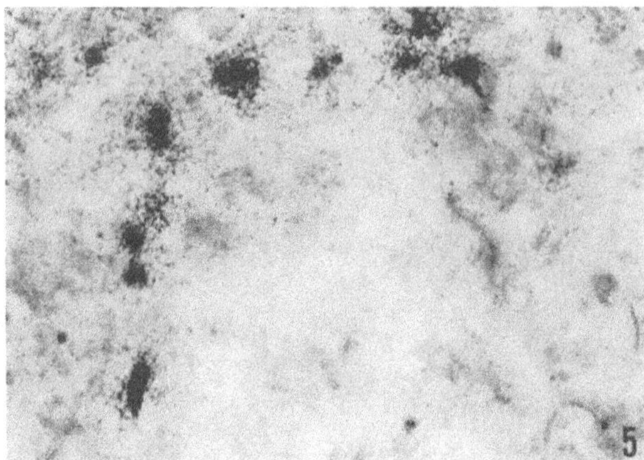

Fig. 5. Autoradiograph of a sectioned spinal ganglion processed after "reactivation" of latent Herpes simplex virus. The ganglion had been maintained for three days in a millipore chamber and placed intraperitoneally in a noninfected recipient mouse. After sectioning, the tissue was processed by *in situ* hybridization methods to detect Herpes simplex virus specific DNA. Note multiple sites of reactivation demonstrated by the presence of silver grains. ×300

from latently infected ganglia taken from the animal and assayed directly. However, when these ganglia were transplanted to uninfected syngeneic mice, viral antigens could easily be detected in *widely scattered* neurons after 2–3 days of incubation (STEVENS and COOK, 1974). In addition, no viral products could be seen when ultrastructural methods were used to examine latently infected ganglia, and no viral antigens are seen in sections of latently infected ganglia stained by immunofluorescent methods using antisera prepared against all viral-induced antigens (STEVENS and COOK, 1971). Finally, when *in situ* nucleic acid hybridization experiments were performed on latently infected ganglia only rarely (≤ 0.05 neurons/ganglionic section) are neurons found which contain viral DNA (STEVENS and COOK, 1974). If such ganglia are transplanted to uninfected mice, within 3 days, 5–10 *widely scattered* neurons/ section are found to contain such DNA (Fig. 5).

Although each of these observations is, in itself, not unequivocal evidence against the dynamic state hypothesis, taken together, they represent considerable support for the static state alternative. In this regard, the last two are the most impressive since they indicate that a significant number of neurons possess viral genomic material which cannot be detected either directly (by nucleic acid hybridization methods) or indirectly (after expression as antigens) in ganglia taken directly from the latently infected animal. Finally, one piece of information could be taken as support for the dynamic state hypothesis. When searching exhaustively in latently infected rabbit trigeminal ganglia, BARINGER and SWOVELAND (1974) found a rare neuron in which viral replication was taking place. The very few neurons mentioned above in which we have detected viral DNA in latently infected spinal ganglia is probably the murine counterpart of this finding. Although this result seems to support the "dynamic state" hypothesis, at the moment it is equally likely that these observations represent "spontaneous" reactivations of active infection in a very few latently infected neurons. As ROIZMAN (1974) quite correctly points out, this important and fundamental problem is unlikely to be clearly resolved until a more simple system which can be manipulated *in vitro* is developed.

C. Concerning the Mechanism by which the Latent Infection is Maintained

The finding that, after *in vitro* cultivation or transplantation, multiple and widely separated neurons are "induced" to replicate viral DNA and synthesize viral-specific antigens is significant in another context. It clearly indicates that a means of repressing viral replication or inactivating infectious virus which operated *in vivo* is lost when the ganglia are maintained *in vitro*. There is now a growing body of evidence that cells harboring viral genomic material produce factors which inhibit viral replication, and in one herpes-virus system (reviewed by STEVENS and COOK, 1974), anti-viral antibody was suggested to suppress the appearance of viral-specific antigens in cells carrying the viral genome (AOKI et al., 1972). From these considerations, it seemed to us probable that such a factor was operating in the ganglia of latently infected mice, but that it either was not present or its effectiveness was lost when ganglia were transplanted to noninfected mice or maintained in vitro. As is detailed elsewhere (STEVENS and COOK, 1974) we searched for such a factor by transplanting latently infected ganglia to latently infected and noninfected mice. If such a factor were present and circulating in latently infected mice, it might be predicted that the latent infection would be maintained in latently infected ganglia. If there were no such factor, or if the transplantation procedure rendered the ganglia insensitive to its action, then induction of the virus would be expected to occur as it had in noninfected mice. In summary, a series of experiments revealed that the number of neurons in such transplants which produce viral antigens and DNA at 3–4 days after transplantation, was considerably less in latently infected mice than in control mice. Moreover, when ganglia were transplanted in a chamber enclosed by a nitrocellulose filter of 22 mμ pore size, the phenomenon was still reproduced. A systematic analysis ultimately led to the demonstration that the effect could be completely reproduced when anti-viral IgG was administered to noninfected mice receiving the ganglionic transplant. Thus, IgG appeared to inhibit intraneuronal viral DNA and antigen synthesis, thereby restricting the appearance of infectious virus. Although a role for cytotoxic antibody could not be unequivocally ruled out, certain experiments made such an effect seem unlikely. Therefore, it seems that anti-viral IgG may play an important role in maintenance of the latent infection. A model (admittedly very superficial) which is consistent with the data from this and related systems would involve a specific interaction between the viral specific IgG and viral-induced membrane antigens on the surface of neurons. Through some as yet undefined intracellular effector molecules, this interaction could repress the complete expression of the viral genome.

Although these findings do not appear to easily "fit" with the observation that individuals with recurrent herpetic disease possess antiviral antibody, they are certainly not incompatible. For example, the provocations that are known to induce recurrent disease in man may overcome the effect of antibody just as transplantation to latently infected or immune mice overcomes the effect of antibody in the few neurons which are induced even in this situation. These considerations are discussed in greater detail elsewhere (STEVENS and COOK, 1974)[4].

VI. Reactivation of Active Infection

From the considerations presented, it is clear that, following peripheral inoculation, HSV does produce latent infections in sensory ganglia in both animals and man. In addition, it is now generally accepted that the virus uses sensory nerves as the means to travel between the sites. It is also probable that the virus travels in axons and remains latent in neurons in a non-replicating state which may be mediated by anti-viral IgG. However, a knowledge of the mechanisms operative in reactivation is obviously of great importance, and this phenomenon is still undefined. It is well known that diverse stimuli reactivate herpetic infections in man. For example, menstruation, emotional disturbances, exposure to ultraviolet light, fever, and trigeminal root section all have been reported to induce herpetic disease. Active herpes simplex infections have been reported to be complications of immunosuppressive measures used in transplantation procedures, and corticosteroids are known to aggravate both natural and experimental herpetic ocular disease (reviewed by STEVENS and COOK, 1973). Experimentally, rabbits have been shown to reactivate ocular infection spontaneously and after epinephrine stimulation (LAIBSON and KIBRICK, 1966; ANDERSON et al., 1961), and some years ago, previously infected animals were found to develop encephalitis after immunologic or pharmacologic manipulation (GOOD and CAMPBELL, 1948; SCHMIDT and RASMUSSEN, 1960). From this, it would seem that rabbits would be desirable experimental animals in which to study reactivation. However, the expense involved in such studies makes a search for other systems desirable. To this end, we spent a great deal of time and

---

4 It is important to also consider the role of cell-mediated immunity in recurrent herpetic disease. Several reports have shown that such specific immunity exists in both man and experimental animals (see review by NAHMIAS and ROIZMAN, 1973), and one report (WILTON et al., 1972) suggests that certain specific functions ascribed to lymphocytes may be deficient in individuals with recurrent disease. Although our experiments indicate that circulating antibodies are of crucial importance in maintenance of the latent state in spinal ganglia, it seems a certainty that specific cell-mediated immunity (in conjunction with humoral immunity) plays a crucial role in arresting development of the lesion at the peripheral site in skin, mucous membrane, or eye after virus has been activated.

effort attempting to develop a model for reactivation in the mouse. As is detailed elsewhere (STEVENS and COOK, 1973 b), until recently attempts to reactivate virus and then reprecipitate neurologic or cutaneous disease in mice harboring latent virus in sacrosciatic spinal ganglia have not been encouraging. In addition, unlike the system in rabbits, we were unable to demonstrate spontaneous ocular infections in mice harboring latent virus in trigeminal ganglia (KNOTTS et al., 1974). Recently, however, encouraging results derived from latently infected mice have been obtained in two laboratories. Remembering the impressive observations that cutaneous herpetic disease follows trigeminal root section (WALZ et al., 1974) established latent infections in the lumbosacral spinal ganglia of mice, and then sectioned branches of the sciatic nerve at the point at which they emerged from the intervertebral foramina. Three days later, in about 1/3 of the animals, infectious virus could be recovered when the corresponding ganglia (but not the ganglia from control animals) were ground and assayed directly for infectious virus.

More recently, we have shown that pneumococcal pneumonitis results in reactivation of infectious virus in sacro-sciatic spinal ganglia. In these experiments (STEVENS et al., manuscript in preparation) 5×10^7 bacteria were given intratracheally. Twenty-four hours later, the mice became gravely ill, but survived when intensive penicillin therapy was instituted. Beginning on the day the experiment was begun and continuing for the next 5 days, several mice were sacrificed daily. The spinal ganglia were ground and assayed directly for virus, and corresponding sciatic nerves and roots were cocultivated with susceptible cells *in vitro*. It was found that infectious virus could be recovered from ganglia at 1, 2, and 3 days after bacterial infection. In addition, from a background level of about 10% of mice at day 0, there was an increasing number of animals from which cocultivation methods demonstrated virus in the nerves, and a maximum of 40% was reached on the 4th day. Thus, after bacterial infection, infectious virus was induced in ganglia, and cocultivation techniques revealed the presence of virus in the associated nerves. The sequence of appearance of virus in the ganglia, followed by appearance of infectious virus in the nerves suggests strongly that the virus present in nerves came from the ganglia and is therefore traveling centrifugally in the nerve. It should also be noted that we have never found virus in the feet of these mice, nor have cutaneous lesions or neurologic signs developed.

Although neither of these models is the ideal one since clinically apparent disease does not follow the "induction", they are encouraging and make further attempts to develop a more instructive system in the mouse justifiable.

VII. Latent Infections in the Central Nervous System

Although this review is primarily concerned with latent infections in sensory ganglia and the role of these organs in the pathogenesis of recurrent cutaneous disease, a brief consideration of latent infections in the central nervous system seems appropriate. In man, at least some cases of herpetic encephalitis have been suggested to result from reactivation of a latent infection (cf. Leider et al., 1965; Kibrick and Gooding, 1965). Some 30 years ago, Good and Campbell (1948) succeeded (by inducing anaphylactic shock in the animal) in inducing encephalitis in rabbits previously infected with HSV. Later, Schmidt and Rasmussen (1960) accomplished the same result with epinephrine. These results indicated that the agent remained in a persistent or latent state in the brain (in both experiments infectious virus was present in the brains of control as well as encephalitic animals). More recently, we showed directly that HSV can induce a latent infection in the central nervous system of experimental animals (Knotts et al., 1973). Here, employing the culture techniques described earlier, we showed that latent virus was harbored for months in the brainstems of rabbits and in the spinal cords of mice which had recovered from acute infection of the CNS. These results satisfy a critical, basic event which is required for support of the concept that the nervous system is the source of virus for at least some cases of herpetic encephalitis. It would now be of interest to know if similar results can be derived from human tissues, and whether the experimental system can be further developed and exploited as a model for chronic, degenerative diseases of the nervous system. These results are also consistent with a provocative paper suggesting that a correlation existed between high antibody titers to HSV-1 and aggressive psychopathic states (Cleobury et al., 1971). At this juncture it may not be too fanciful to at least consider the possibility that some psychiatric illnesses may result from latent or reactivated herpetic disease in appropriate areas of the central nervous system.

VIII. The Future

From the experimental and clinical evidence reviewed here, it can be concluded that Herpes simplex virus remains latent in sensory ganglia and travels in nerves. It is also likely that the endoneural route taken is axonal. In addition, the virus probably persists in some nonreplicating state in neurons, and this interaction may be modulated by specific immune IgG. Although a reactivation of infectious virus following physical and infectious manipulation has been achieved in murine systems, the biochemical basis for this observation is completely undefined, and, except for the central nerveous system disease of rabbits, reproducible induction of a clinically apparent disease has yet to be achieved in any experimental system. In addition, the state of the viral genome in latently infected neurons has yet to be defined.

Because of the recent successes in reactivation of active infection, in the immediate future it seems quite likely that a reproducible and predictable clinical disease associated with particular latently infected ganglia will be produced in some experimental system. If, following this demonstration, the disease can be prevented by interruption of appropriate nerve trunks, the last significant obstacle for acceptance of the general hypothesis presented earlier will have been removed.

A resolution of the other two unknowns remaining (state of the virus in neurons, and biochemical basis of latency) seem to be further away. Employing nucleic acid hybridization methods on nitrocellulose membranes, we have been unable to detect the viral genome in latently infected ganglia [sensitivity = 10–20 genome equivalents/cell (STEVENS and COOK, 1973 b)]. However it is possible that the application of the more sensitive techniques presently available (cf. FRENKEL et al., 1972; HUANG et al., 1973; ADAMS et al., 1973) will allow a determination of whether or not the viral genome is integrated into the host cell chromosome. If integration is shown to occur, strong (but not unequivocal) direct evidence for maintenance of the virus as a non-replicating agent will have been derived. A definition of the biochemical basis of latency (the intracellular phenomena which follow the interaction of infected neurons with antiviral antibody) can be seriously investigated when a system involving a uniform population of cells (neurons) all harboring or potentially able to harbor latent virus can be studied. In this regard, we are now attempting by various manipulations both *in vivo* and *in vitro*, to establish latent infections in the mouse C-1300 neuroblastoma line. The system is particularly attractive since the cells can be successfully transplanted to a syngeneic host which has been manipulated by virologic, immunologic, and pharmacologic means.

Finally, the role of reactivated HSV in the genesis of acute herpetic encephalitis, and the possible serious, long-standing complications of latent infections in the central nervous system can, and should be systematically investigated.

Acknowledgements. In this communication, I have drawn upon appropriate information gathered and synthesized by several previous reviewers. I would particularly recognize those reviews written by C. E. VON ROYEN and A. J. RHODES (1948); T. F. PAINE, JR. (1964), B. ROIZMAN (1965, 1974), and P. WILDY (1973). In addition, this laboratory is especially grateful to R. W. SCHLESINGER whose encouragement at a critical time was greatly responsible for our decision to investigate the pathogenesis of herpetic infections. The research from which our own data were derived was supported by The National Institutes of Health, United States Public Health Service (Grants AI-06246 and NS-08711).

References

ADAMS, A., LINDAHL, T., KLEIN, G.: Linear association between cell DNA and Epstein-Barr virus DNA in a human lymphoblastoid cell line. Proc. nat. Acad. Sci. (Wash.) **70**, 2888–2895 (1973)

ANDERSON, W. A., MAGRUDER, B., KILBOURNE, E. D.: Induced reactivation of Herpes simplex virus in healed rabbit corneal lesions. Proc. Soc. exp. Biol. (N.Y.) **107**, 628–631 (1961)

ANDREWES, C. H., CARMICHAEL, E. A.: A note on the presence of antibodies to herpes virus in post-encephalitic and other human sera. Lancet **1**, 857–858 (1930)

AOKI, I., GEERING, G., BETH, E., OLD, L. J.: In: Recent advances in human tumor virology and immunology. W. NAKAHARA, K. NISHIOKA, T. HIRAYOMA, Y. ITO (eds.). Tokyo: Univ. of Tokyo Press 1972

BARINGER, J. R.: Recovery of Herpes simplex virus from human facial ganglion. Program, American Assoc. of Neuropathologists 20 (1974)

BARINGER, J. R., SWOVELAND, P.: Recovery of Herpes simplex virus from human trigeminal ganglia. New Engl. J. Med. **288**, 648–650 (1973)

BARINGER, J. R., SWOVELAND, P.: Persistent Herpes simplex virus infection in rabbit trigeminal ganglia. Lab. Invest. **30**, 230–240 (1974)

BASTIAN, F. O., RABSON, A. S., YEE, C. L., TRALKA, T. S.: Herpesvirus Hominis: isolation from human trigeminal ganglion. Science **178**, 306–307 (1972)

BURNET, F. M., WILLIAMS, S. W.: Herpes simplex: a new point of view. Med. J. Aust. **1**, 637–642 (1939)

CARTON, C. A.: Effect of previous sensory loss on the appearance of herpes simplex. J. Neurosurg. **10**, 463–468 (1953)

CARTON, C. A., KILBOURNE, E. D.: Activation of latent herpes simplex by trigeminal sensory-root section. New Engl. J. Med. **246**, 172–176 (1952)

CLEOBURY, J. F., SKINNER, G. R. B., THOULESS, M. E., WILDY, P.: Association between psychopathic disorder and serum antibody to Herpes simplex virus (Type 1). Brit. med. J. **1**, 438–439

COOK, M. L., BASTONE, V. B., STEVENS, J. G.: Evidence that neurons harbor latent Herpes simplex virus. Inf. and Imm. **9**, 946–951 (1974)

COOK, M. L., STEVENS, J. G.: Pathogenesis of herpetic neuritis and ganglionitis in mice: evidence for intra-axonal transport of infection. Inf. and Imm. **7**, 272–288 (1973)

CORIELL, L. L.: Discussion of paper by B. ROIZMAN. In: Virus, nucleic acids and cancer. Baltimore: Williams & Wilkins Co. 1963

CUSHING, H.: Perineal zoster. Bull. Johns Hopk. Hosp. **158**, 172 (1904)

CUSHING, H.: The surgical aspects of major neuralgia of the trigeminal nerve. A report of 20 cases of operation on the gasserian ganglion, with anatomic and physiologic notes on the consequences of its removal. J. Amer. med. Ass. **44**, 773–779, 860–865, 920–929, 1002–1008, 1088–1093 (1905)

DILLARD, S. H., CHEATHAM, W. J., MOSES, H. L.: Electron microscopy of zosteriform herpes simplex infection in the mouse. Lab. Invest. **26**, 391–402 (1972)

DODD, K., JOHNSTON, L. M., BUDDINGH, G. J.: Herpetic stomatitis. J. Pediat. **12**, 95–102 (1938)

DOERR, R.: Sitzungsberichte der Gesellschaft der schweizerischen Augenärzte. Diskussion. Klin. Mbl. Augenheilk. **65**, 104 (1920)

DOERR, R., SCHNABEL, A.: Das Virus des Herpes febrilis und seine Beziehungen zum Virus der Encephalitis epidemica (lethargica). Z. Hyg. Infekt.-Kr. **94**, 29–81 (1921)

DOERR, R., VÖCHTING, K.: Etudes sur le virus de l'herpes febrile. Rev. gén. Ophtal. (Paris) **34**, 409–421 (1920)

DOUGLAS, R. G., COUCH, R. B.: A prospective study of chronic Herpes simplex virus infection and recurrent herpes labialis in humans. J. Immunol. **104**, 289–295 (1970)

FINDLAY, G. M., MacCALLUM, F. O.: Recurrent traumatic herpes. Lancet **238**, 259–261 (1940)

FRENKEL, N., ROIZMAN, B., CASSAI, E., NAHMIAS, A.: A DNA fragment of herpes simplex 2 and its transcription in human cervical cancer tissue. Proc. nat. Acad. Sci. (Wash.) **12**, 3784–3789 (1972)

GOOD, R. A., CAMPBELL, B.: The precipitation of latent herpes simplex encephalitis by anaphylactic shock. Proc. Soc. exp. Biol. (N.Y.) **68**, 82–87 (1948)

GOODPASTURE, E. W.: The axis cylinders of peripheral nerves as portals of entry to the central nervous system for the virus of herpes simplex in experimentally infected rabbits. Amer. J. Path. **1**, 11–28 (1925a)

GOODPASTURE, E. W.: The pathways of infection of the central nervous system in herpetic encephalitis of rabbits contracted by contact, with a comparative comment on medullary lesions in a case of human poliomyelitis. Amer. J. Path. **1**, 29–46 (1925b)

GOODPASTURE, E. W.: Herpetic infections with special reference to involvement of the nervous system. Medicine (Baltimore) **7**, 223–243 (1929)

GOODPASTURE, E. W., TEAGUE, O.: Transmission of the virus of herpes fibrilis along nerves in experimentally infected rabbits. J. med. Res. **44**, 139–184 (1923)

GRÜTER, W.: Experimentelle und klinische Untersuchungen über den sog. Herpes corneae. Klin. Mbl. Augenheilk. **65**, 398–399 (1920)

HEAD, H., CAMPBELL, A. W.: The pathology of herpes zoster and its bearing on sensory localization. Brain **23**, Part III, 353–523 (1900)

HOWARD, W. T.: The pathology of labial and nasal herpes and of herpes of the body occurring in acute croupous pneumonia and their relation to so-called herpes zoster. Amer. J. med. Sci. **125**, 256–272 (1903)

HOWARD, W. T.: Further observations on the relation of lesions of the gasserian and posterior root ganglia to herpes occurring in pneumonia and cerebrospinal meningitis. Amer. J. med. Sci. **130**, 1012–1019 (1905)

HUANG, E.-S., CHEN, S.-T., PAGANO, J. S.: Human cytomegalovirus I purification and characterization of viral DNA. J. Virol. **12**, 1473–1481 (1973)

JOHNSON, R. T.: The pathogenesis of herpes virus encephalitis I. virus pathways to the nervous system of suckling mice demonstrated by fluorescent antibody staining. J. exp. Med. **119**, 343–356 (1964)

KAUFMAN, H. E., BROWN, D. C., ELLISON, E. M.: Recurrent herpes in rabbit and man. Science **156**, 1628–1629 (1967)

KIBRICK, S., GOODING, G. W.: Pathogenesis of infection with Herpes simplex virus with special reference to nervous tissue. In: Slow, latent and temperate virus infections. GADJUSEK, D. C., GIBBS, C. J., ALPERS, M. (eds.) NINDB. Monogr. No. 2. Washington, D. C.: U.S. Gov't Printing Office 1965

KNOTTS, F. B., COOK, M. L., STEVENS, J. G.: Latent Herpes simplex virus in the central nervous system of rabbits and mice. J. exp. Med. **138**, 740–744 (1973)

KNOTTS, F. B., COOK, M. L., STEVENS, J. G.: Pathogenesis of herpetic encephalitis in mice following ophthalmic inoculation. J. infect. Dis. **130**, 16–27 (1974)

KRISTENSSON, K., LYCKE, E., SJÖSTRAND, J.: Spread of Herpes simplex virus in peripheral nerves. Acta neuropath. (Berl.) **17**, 44–53 (1971)

LAIBSON, P. R., KIBRICK, S.: Reactivation of herpetic keratitis by epinephine in rabbit. Arch. Ophthal. **75**, 254–260 (1966)

LAVAIL, J. H., LAVAIL, M. M.: Retrograde axonal transport in the central nervous system. Science **176**, 146–147 (1972)

LEIDER, W., MAGOFFIN, R. L., LENNETE, E. H., LEONARDS, L. M. R.: Herpes simplex-virus encephalitis: its possible association with reactivated latent infection. New Engl. J. Med. **273**, 341–347 (1965)

LEVADITI, C., HARVIER, P.: Etude experimentale de l'encephalite dite lethargique. Ann. Inst. Pasteur **34**, 911–972 (1920)

LÖWENSTEIN, A.: Ätiologische Untersuchungen über den fieberhaften Herpes. Münch. med. Wschr. **66**, 769–770 (1919)

MARINESCO, G., DRAGENESCO, S.: Recherches experimentales sur le neurotropisme du virus herpetique. Ann. Inst. Pasteur **37**, 753–783 (1923)

NAHMIAS, A. J., ROIZMAN, B.: Infection with Herpes simplex virus 1 and 2. New Engl. J. Med. **289**, 667–674, 719–724, 781–789 (1973)

NICOLAU, S., POINCLOUX, P.: Resultats d'une greffe de plave saine faite in 1924 sur un doiget atteint d'herpes recidivant perdant les cinq. annes precedentes. C.R. Soc. Biol. (Paris) **98**, 368 (1928)

PAINE, T. F., JR.: Latent herpes simplex infection in man. Bact. Rev. **28**, 472–479 (1964)

RODDA, S., JACK, I., WHITE, D. O.: Herpes simplex virus from trigeminal ganglion. Lancet **1**, 1394 (1973)

ROIZMAN, B.: An inquiry into the mechanisms of recurrent herpes infections of man. Perspectives in virology, vol. IV, M. POLLARD (ed.). New York: Hoeber 1964

ROIZMAN, B.: Herpesvirus, latency and cancer: a biochemical approach. R.E.S. **7**, 312–320 (1974)

RUSTIGIAN, R., SMULOU, J. B., TYE, M., GIBSON, W. A., SHINDELL, E.: Studies on latent infection of skin and oral mucosa in individuals with recurrent herpes simplex. J. invest. Derm. **47**, 218–221 (1966)

SCHMIDT, J. R., RASMUSSEN, A. F., JR.: Activation of latent herpes simplex encephalitis by chemical means. J. infect. Dis. **106**, 154–158 (1960)

SCHWARTZ, J., ELIZAN, T. S.: Chronic herpes simplex infection. Arch. Neurol. (Chic.) **28**, 224–230 (1973)

SCOTT, T. F. M.: Epidemiology of herpetic infections. Amer. J. Ophthal. **43**, 134–143 (1957)

Scott, T. F. M., Tokumaru, T.: The herpesvirus group. In: Viral and rickettsial infections of man. 4th ed. Horsfall, F., Tamm, I. (eds.). Philadelphia: Lippincott 1965

Stalder, W., Zurukzoglu S.: Experimentelle Untersuchungen über Herpes. Transplantation herpesinfizierter Hautstellen, Reaktivierung von abgeheilten, künstlich infizierten Hautstellen, Herpesbehandlung. VI. Mitt. Zbl. Bakt. **136**, 94–97 (1936)

Stevens, J. G., Cook, M. L.: Latent Herpes simplex virus in spinal ganglia of mice. Science **173**, 843–845 (1971)

Stevens, J. G., Cook, M. L.: Latent Herpes simplex virus in sensory ganglia. In: Perspectives in virology, vol. VIII, M. Pollard (ed.). New York: Academic Press 1973a

Stevens, J. G., Cook, M. L.: Latent herpes simplex infection. In: Virus research, C. F. Fox and W. S. Robinson (eds.). New York: Academic Press 1973b

Stevens, J. G., Cook, M. L.: Maintenance of latent herpetic infection: an apparent role for anti-viral IgG. J. Immunol. **113**, 1685–1693 (1974)

Stevens, J. G., Nesburn, A. B., Cook, M. L.: Latent Herpes simplex virus from trigeminal ganglia of rabbits with recurrent eye infection. Nature (Lond.) New Biol. **235**, 216–217 (1972)

van Rooyen, C. E., Rhodes, A. J.: Herpes febrilis. In: Virus diseases of man, 2nd ed. New York: T. Nelson & Sons 1948

Vidal, E.: Inoculabilite des pustules d'ecthyma. Ann. Derm. Syphiligr. **4**, 350–358 (1873)

Walz, M. A., Price, R. W., Notkins, A. L.: Latent infection with Herpes simplex virus types 1 and 2: viral reactivation *in vivo* after neurectomy. Science **184**, 1185–1187 (1974)

Wildy, P.: The progression of Herpes simplex virus to the central nervous system of the mouse. J. Hyg. (Lond.) **65**, 173–192 (1967)

Wildy, P.: Herpes: history and classification. In: The herpesviruses. Kaplan, A. S. (ed.). New York: Academic Press 1973

Wilton, J. M. A., Ivanyi, L., Lehner, T.: Cell-mediated immunity in herpesvirus hominis infections. Brit. med. J. **1**, 723–726 (1972)

Zalená, J.: Bidirectional shift of mitochondria in axons after injury. In: Cellular dynamics of the neuron, Symposium of the International Society for Cell Biology, S. H. Barondes (ed.), vol. 18. New York: Academic Press 1969

Temperature-Sensitive Mutants of Herpesviruses

P. A. SCHAFFER[1]

With 5 Figures

Table of Contents

1 Dept. of Virology and Epidemiology, Baylor College of Medicine, Texas Medical Center, Houston, Texas 77025, U.S.A.

I. Introduction

The processes involved in herpesvirus replication, latency, and oncogenic transformation, have, in general, been rather poorly defined. A primary reason for this is the size and complexity of the herpesvirus genome. Undoubtedly, a better understanding of the functions of the viral genome in infected and transformed cells will be achieved through studies with temperature-sensitive (*ts*) mutants of herpesviruses since, theoretically, any essential gene function can be affected by mutants of this type.

A. The Herpesviruses

A consideration of the genetic analysis of members of the herpesvirus group necessitates a description, albeit brief, of the properties of the group and, most importantly, of their genetic material. The herpesviruses comprise a group of relatively large (100–150 nm), enveloped viruses. The envelope surrounds an icosahedral capsid enclosing a core which contains double-stranded DNA (Roizman, 1969). The group is thus defined on the basis of a common virion morphology. In addition to a common structure, members of the group share a number of biological properties such as a similar replicative cycle, the ability to cause latent and chronic infections, and the ability to induce antigenic modifications of infected cell membranes. Several herpesviruses have been associated recently with malignancies in man and animals (Klein, 1972). Herpesviruses are ubiquitous and have been described in over 30 different species (Hunt and Melendez, 1969; Wildy, 1971; Farley et al., 1972; Kazama and Schornstein, 1972; Nahmias et al., 1972; Roizman et al., 1973). Their widespread occurrence in nature suggests a common ancestor. Although their antigenic relationships have not been thoroughly investigated, members of the group, for the most part, appear to be antigenically distinct. However, common antigens have been described among herpesviruses which infect different species (Kirkwood et al., 1972; Ross et al., 1972a; Wildy, 1973). Most notably, antigenic relationships have been demonstrated between herpes simplex virus (HSV) type 1 of man, B virus of Old-World monkeys, and pseudorabies virus (PRV) of pigs (Watson et al., 1967), and between Marek's disease virus (MDV) of chickens and turkey herpesvirus (Witter et al., 1970).

Although all herpesviruses possess the same virion structure and exhibit similar biological properties, the characteristics of their DNAs vary considerably. The base compositions of herpesvirus DNAs range from 33 moles percent guanine plus cytosine (G + C) (canine herpesvirus) to 72 moles percent G + C (infectious bovine rhinotracheitis virus, IBR) (Plummer et al., 1969). The range of G + C content of the DNAs of different herpesviruses is a striking feature of the group in that it is greater than that of any other virus group, or indeed, of any known biological group (Wildy, 1973). The implications of such base sequence heterogeneity for the comparative genetic analysis of these viruses are obviously far reaching. The molecular weights of the DNAs of several herpesviruses have been estimated by a variety of methods and range from 54×10^6 to 103×10^6 daltons (RUSSELL and CRAWFORD, 1964; BECKER et al., 1968; BACHENHEIMER et al., 1972; GRAHAM et al., 1972; WAGNER et al., 1974). The heterogeneity of herpesvirus DNAs has been demonstrated by two additional pieces of evidence. Firstly, with the exception of HSV type 1 and 2 DNAs, which share 50–70% base-sequence homology (KIEFF et al., 1972; LUDWIG et al., 1972), no significant homology has been demonstrated between the DNAs of HSV, MDV, Epstein-Barr virus (EBV), or human and non-human cytomegaloviruses (CMV) using DNA:DNA hybridization techniques (ZUR HAUSEN and SCHULTE-HOLTHAUSEN, 1970; BACHENHEIMER et al., 1972; HUANG and PAGANO, 1974). However, LUDWIG et al. (1972) reported an 8–10% complementarity between PRV-DNA and HSV type 1 and 2 DNAs. Secondly, SUBAK-SHARPE (1967) demonstrated profound differences in the general designs (base-doublet patterns) of the DNAs of HSV, PRV, and equine abortion virus (EAV) using the technique of nearest-neighbor analysis. Assuming a common ancestor, this would suggest a considerable degree of evolutionary divergence. It is possible that divergent evolution due to a series of neutral mutations scattered throughout the viral genome may have resulted in an alteration of herpesvirus DNA which markedly affected the antigenic composition of viral polypeptides with little effect on their structural and functional properties (FITCH, 1972). In this manner, base sequences may have evolved to such a degree that no stable hybrids can be observed between herpesvirus DNAs using available techniques.

If, for purposes of illustration, studies of HSV-DNA are extrapolated to the DNAs of other herpesviruses, several features of the composition and configuration of HSV DNA are of importance in a consideration of the genetics of herpesviruses. (1) Evidence has been obtained that the HSV DNA molecule has single-strand interruptions or alkali-labile bonds in one or both strands of the linear duplex (FRENKEL and ROIZMAN, 1972; GORDIN et al., 1973; WILKIE, 1973; BISWAL et al., 1974; HIRSCH and VONKA, 1974). It is possible that such labile interruptions may represent preferred sites of breakage and reunion during genetic recombination as suggested by LANNI et al. (1966) for T5. (2) The DNA molecules of EAV (SOEHNER et al., 1965) and HSV (BECKER et al., 1968; BACHENHEIMER et al., 1972) are presently thought to be linear duplexes. However, WILKIE (1973) has suggested the intriguing possibility that HSV DNA, like that of T2, may be a circularly permuted, terminally redundant collection of linear duplexes. Indeed, terminal redundancy in HSV-DNA has been demonstrated only recently by GRAFSTROM (personal communication) and SHELDRICK (personal communication). (3) By inference, from the work of ERIKSON and SZYBALSKI (1964), HSV-DNA is not glycosylated, nor apparently are HSV and PRV DNAs methylated (LOW et al., 1969, 1971). (4) Unusual bases have not been reported for herpesvirus DNAs, and (5) the DNA of a least one member of the group, HSV, is apparently both infectious (GRAHAM et al., 1973; SCHELDRICK et al., 1973) and capable of inducing transformation *in vitro* (WILKIE, personal communication).

With regard to the coding capacity of herpesvirus DNA, we again turn to the information available on the DNA of HSV. With a molecular weight of about 100×10^6 daltons, HSV types 1 and 2 possess sufficient genetic information to code for about 50000 amino acids, corresponding to 70–80 polypeptides with an average molecular weight of 85000 daltons (SPEAR and ROIZMAN, 1972). Unlike the DNA of HSV type 1, however, about 16% of the DNA of HSV type 2 appears to consist of repetitive sequences (ROIZMAN and FRENKEL, 1973). Employing high resolution polyacrylamide gel electrophoresis, HONESS and ROIZMAN (1973) reported the presence in HSV type 1-infected cells of at least 47 virus-specific polypeptides with a combined molecular weight corresponding to 41000 amino acids, which accounts for about 75% of the coding capacity of the viral DNA. Twenty-four of these polypeptides have been identified as structural components of the virion (SPEAR and ROIZMAN, 1972) and 16 as non-structural components; the identity of the remaining polypeptides is still uncertain (HONESS and ROIZMAN, 1973).

The herpesviruses thus present a complex problem for genetic analysis. Not only are their genomes large but, despite the extensive structural and functional similarities among members of the group, their DNAs differ markedly in both base composition and general design. A sensitive means for examining the extent of the functional and genetic relatedness of these viruses is now available through the use of *ts* mutants.

B. Temperature-Sensitive (*ts*) Mutants

Ts mutants are able to replicate at low, permissive temperature (pT°) but not at high nonpermissive temperature (npT°). At the molecular level, *ts* mutants appear to be defective in replication because an altered base sequence specifies an altered amino-acid sequence in an essential viral protein which renders that protein incapable of assuming or maintaining a functional configuration at the npT°. Mutant replication occurs normally, or nearly so, at pT°, resulting in the production of infectious progeny virus which still carries the potentially lethal mutation. The expression of *ts* mutations thus depends upon a fundamental property which is common to all proteins (i.e., susceptibility to heat inactivation) rather than on a specific function. Mutants defective in a given protein can thus be isolated even though nothing is known of that protein's specific function. Advantages of *ts* mutants for genetic analysis include the following: (1) The isolation of sufficient numbers of independent *ts* mutants for a given virus can ultimately lead to the identification of most essential gene functions as demonstrated in bacterial virus systems. (2) *Ts* mutations are not limited to genes which determine modifiable characters such as plaque type and host range, but they can occur in all genes with either essential or nonessential functions. Thus, these mutants theoretically permit the complete genetic mapping of a virus. (3) Complementation tests can lead to the identification of different genes at the functional level. (4) Essential functions can be analyzed by determining the physiological nature of the block in mutant replication under restrictive conditions.

The use of *ts* mutants in the genetic analysis of a number of animal viruses has been reviewed extensively by FENNER (1969) and GHENDON (1972). The present review will consider the work which has been done to date with *ts* mutants of herpesviruses.

II. Induction, Isolation, and Growth Properties of *ts* Mutants of Herpesviruses

A. Induction

Although the most extensive genetic analyses of herpesviruses have been conducted with *ts* mutants of HSV types 1 and 2, *ts* mutants of PRV, IBR, HVS, and MDV have also been isolated. The origin and growth properties of these mutants are presented in Table 1.

A variety of cell types have been utilized in studies of herpesvirus *ts* mutants and pT° and npT° have ranged from 30°–36° and 37.5°–41°, respectively (Table 1). The selection of the cell type and permissive and restrictive temperatures for use in these studies was an obvious consequence of the particular herpesvirus and wild-type (WT) virus strain studied. The innate thermal sensitivity of herpesviruses complicates studies with *ts* mutants since thermal inactivation of both WT virus and *ts* mutant progeny at the npT° results in an under-estimate of the actual yield of infectious virus. The use of different temperatures and host cells in comparative genetic studies with different sets of mutants will no doubt complicate these studies since: (a) the optimum temperatures at which different sets of mutants replicate may vary significantly, and (b) it is possible that certain herpesvirus *ts* mutants may be temperature-sensitive in one host-cell system but not in another, as in the case of the HSV type 2 *ts* mutants of KOMENT and RAPP (1974).

Ts mutants have been derived from different WT strains of HSV types 1 and 2, a fact which should prove to be useful in intratypic comparisons of strains of the two types of HSV. Exchange of mutants derived from different WT strains for complementation and recombination analysis should ultimately help to identify any differences in the structural organization of the genome which may exist among strains of HSV type 1 and among strains of HSV type 2.

In order to ensure a genotypically and phenotypically homogeneous WT-virus population for mutagenesis, plaque-purified viruses were used for the derivation of *ts* mutants of HSV type 1 (SCHAFFER et al., 1970, 1973; BROWN et al., 1973; MANSERVIGI, 1974), HSV type 2 (ESPARZA et al., 1974), PRV (PRINGLE et al., 1973), and HVS (SCHAFFER et al., in preparation). In addition, SCHAFFER et al. (1970, 1973), BROWN et al. (1973), ESPARZA et al. (1974), and MANSERVIGI (1974) passed their respective WT strains of HSV types 1 and 2 at the npT° in order to eliminate pre-existing *ts* mutants and to increase

Table 1. Temperature-sensitive

Virus[a]	Strain	Cell culture	pT° and npT° [b]	Method of mutagenesis[c]
HSV type 1	Glasgow Str. 17	BHK21 (C13)	31°; 38°	Propagation in the presence of 6.25 µg/ml BrdUrd (24 hrs, 31°)
HSV type 1	KOS	HEL	35°; 40°	Propagation in the presence of 0.3–5.0 µg/ml BrdUrd (48 hrs, 33°)
HSV type 1	KOS	HEL	34°; 39°	a) Propagation in the presence of 1.25–2.5 µg/ml BrdUrd (48 hrs, 33°) b) Propagation in the presence of 20 µg/ml NTG (various times, 34°) c) Treatment with UV-irradiation (30 or 60 sec)
HSV type 1	13	HEp-2	35°; 40°	Propagation in the presence of 3–300 µg/ml BrdUrd (48 hrs, 34°)
HSV type 2	HSG52	BHK21 (C13)	31°; 38°	Propagation in the presence of 5 µg/ml BrdUrd (48 hrs, 31°)
HSV type 2	186	HEL	34°; 38°	Propagation in the presence of 0.3–2.5 µg/ml BrdUrd (48 hrs, 34°)
HSV type 2	IPB2	PRK	35°; 37.5*39°	Treatment with HNO_2
HSV type 2	333	HEF	33°; 39°	a) Propagation in the presence of BrdUrd or NTG b) Treatment with UV-irradiation
PRV	PRV	BHK21 (C13)	31° or 37°; 41°	Propagation in the presence of 10 µg/ml FUra, 10 µg/ml FdUrd and 5 µg/ml BrdUrd (24 hrs, 31°) in cells pretreated with 10 µg/ml FUra (18 hrs)
PRV	BUK	CEF	33°; 40°	a) Propagation in the presence of BU b) Treatment with HNO_2
IBR	NRRL 5670	FBK	30°; 38° or 39°	Treatment with HNO_2
HVS	SM # 11	Vero	34°; 39°	Propagation in the presence of 0.15–5.0 µg/ml BrdUrd
MDV	Biken C2	QEF	36°; 41°	Propagation in the presence of 20 µg/ml NTG (5 days, 36°)

mutants of herpesviruses

No. stable ts mutants/ No. plaque isolates tested	Frequency of ts mutant isolation	No. mutants in study	Growth properties of mutants EOP[d]	EOR[e]	References
no info[f]	no info	9	$10^{-3.4}-<10^{-8.7}$	no info	Brown et al., 1973
22/708	3%	22	$10^{-5.9}-10^{-6.2}$	no info	Schaffer et al., 1970
3/476	0.6%				
7/527	1.3%	22	$10^{-4.0}-10^{-7.8}$	$10^{-4.9}-10^{-5.0}$	Schaffer et al., 1973
6/391	1.5%				
6/351	1.7%	6	$10^{-5.5}-<10^{-8.6}$	$<10^{-8.5}-<10^{-8.0}$	Manservigi, 1974
no info	no info	33	$10^{-4.8}-<10^{-8.6}$	no info	Timbury, 1971
8/394	2%	8	$10^{-4.5}-10^{-7.5}$	$10^{-3.8}-10^{-5.7}$	Esparza et al., 1974
no info	no info	3	no info	$<10^{-4.0}$	Zygraich and Huygelen, 1973
no info	no info	4	no info	no info	Koment and Rapp, 1974
10/1031	1%	10	no info	no info	Pringle et al., 1973
no info	no info	9	no info	no info	Huy et al., 1972
no info	no info	1	no info	$10^{-5.0}$	Zygraich et al., 1974
no info	no info	5	$10^{-3.8}-10^{-4.2}$	no info	Schaffer et al., in preparation
1/120	0.8%	1	no info	$10^{-2.0}$	Onoda et al., 1971

the possibility of selecting WT viruses with enhanced potential for growth at higher temperatures. To facilitate the genetic analysis of their Glasgow 17 WT strain of HSV type 1, Brown et al. took advantage of the fact that it can exist in two stable forms: the *syn* form, which produces syncytial plaques, and the *syn+* form, which produces nonsyncytial plaques. The *syn+* form of the WT virus mutates spontaneously at very low frequency to produce a stable variant, *syn*, with easily recognizable syncytial plaque morphology. Mutants were first isolated as *ts*, *syn* and then since *syn* reverts to *syn+* with low frequency, as *ts*, *syn+* revertants. Every *ts* mutant was thus available simultaneously in both *syn* and *syn+* forms in isogenic material. Timbury et al. (1974) recently reported on the isolation of a stable syncytial variant of their HSV type 2 WT virus and attempts to obtain HSV type 2 *ts* mutants in *syn* and *syn+* forms are planned.

Although spontaneous *ts* mutants of herpesviruses (Manservigi, 1974; Purifoy, personal communication) and other animal viruses are known to occur, the frequency of single mutational defects is increased 10-fold or more when mutagens are employed (Cooper, 1964; Burge and Pfefferkorn, 1966a; Fields and Joklik, 1969; Schaffer et al., 1970). Since mutagens differ in both their mechanisms and sites of action, the use of a variety of mutagens to induce *ts* mutants increases the probability of obtaining mutants with defects in different genes. To date, a number of mutagens have been used to induce *ts* mutants of herpesviruses (Table 1).

Table 1 (continued)

[a]	HSV	= herpes simplex virus.
	PRV	= pseudorabies virus.
	IBR	= infectious bovine rhinotracheitis virus.
	HVS	= herpesvirus saimiri.
	MDV	= herpes-type virus isolated from a chicken with Marek's disease.
[b]	BHK21 (C13)	= BHK21 clone 13.
	HEL	= human embryonic lung fibroblasts.
	PRK	= primary rabbit kidney cells.
	HEF	= hamster embryo fibroblasts.
	CEF	= chicken embryo fibroblasts.
	FBK	= fetal bovine kidney.
	QEF	= quail embryo fibroblasts.
[c]	BrdUrd	= 5-bromodeoxyuridine.
	NTG	= N-methyl-N'-nitro-N-nitrosoguanidine.
	HNO_2	= nitrous acid.
	UV	= ultraviolet light.
	FUra	= 5-fluorouracil.
	FdUrd	= 5-fluorodeoxyuridine.
	BU	= 5-bromouracil.

[d] EOP   = efficiency of plating, ratio of plaque-forming ability $\dfrac{npT°}{pT°}$.

[e] EOR   = efficiency of replication, ratio of yield $\dfrac{npT°}{pT°}$.

[f] No info   = no information.

The most widely used mutagenic technique involves propagating WT virus-infected cells in the presence of varying concentrations of the base analog 5-bromodeoxyuridine (BrdUrd), which replaces thymidine quantitatively and causes transitions in replicating DNA (FREESE, 1963). *Ts* mutants of HSV type 1 (SCHAFFER et al., 1970, 1973; BROWN et al., 1973; MANSERVIGI, 1974), HSV type 2 (TIMBURY, 1971; ESPARZA, 1974), PRV (PRINGLE et al., 1973), and HVS (SCHAFFER et al., in preparation) were induced with BrdUrd. A novel method of BrdUrd mutagenesis was employed by PRINGLE et al. (1973) to obtain *ts* mutants of PRV. Prior to infection with the WT virus, cells were first depleted of deoxythymidine using 5-fluorouracil (FUra) to enhance incorporation of BrdUrd into replicating viral DNA. Cells were then infected and maintained in a medium containing FUra, 5-fluorodeoxyuridine (FdUrd), to block the *de novo* synthesis of thymidine, and BrdUrd, the mutagen. Using this method of mutagenesis, 10 stable *ts* mutants were isolated while the frequency of spontaneous *ts* mutants was less than 0.75% following exposure of infected cells to FUra and FdUrd alone. Whether the combined treatment of infected cells with FUra and FdUrd actually enhanced the incorporation and hence the mutagenic effect of BrdUrd, and whether enhanced incorporation is, indeed, of advantage in the induction of single-step *ts* mutations, remain to be determined.

Whether these methods of mutagenesis have induced *primarily* mutants with single defects also remains to be determined since mutants induced with BrdUrd apparently show much stronger "hot-spotting" than do mutants induced by other agents (DRAKE, 1970). The frequency of *ts* mutant isolation following BrdUrd mutagenesis was 3% or less. The low frequency of mutant isolation obtained in most studies thus suggests that over-mutagenesis and the subsequent isolation of mutants with multiple defects were minimized. However, both HSV type 1 (SCHAFFER et al., 1971) and HSV type 2 (TIMBURY, 1971) mutants which possess multiple *ts* defects have been isolated following mutagenesis with 1.25 and 5 μg/ml of BrdUrd, respectively (the most commonly used concentrations, Table 1). In this regard, therefore, it is likely that other groups of *ts* mutants induced with still higher concentrations of BrdUrd (MANSERVIGI, 1974) will also prove to possess multiple defects (Table 1). Undoubtedly, the problems of over-mutagenesis and multiply defective mutants will complicate the genetic analysis of herpesviruses as they have studies of other animal viruses (COOPER, 1969).

N-methyl-N'-nitro-N-nitrosoguanidine (NTG) has been used to induce *ts* mutants of HSV type 1 (SCHAFFER et al., 1973), HSV type 2 (KOMENT and RAPP, 1974), and MDV (ONODA et al., 1971). Although the precise mechanism of action of NTG is unknown, its presumed mutagenic effect is the methylation of guanine which causes transitions, transversions, and deletions during DNA replication (DRAKE, 1970). The replication point of DNA has been shown to be preferentially mutagenized (CERDÁ-OLMEDO, 1968). In one study (SCHAFFER et al., 1973), the highest frequency of mutant isolation and the greatest rate of virus inactivation occurred when NTG was added to infected cultures at the time of maximum viral DNA synthesis (Roizman, 1969). These observations thus support the concept that NTG acts preferentially on replicating viral DNA.

Ts mutants of HSV type 1 (Schaffer et al., 1973) and type 2 (Koment and Rapp, 1974) have been induced following ultraviolet (UV) irradiation of WT virus. UV irradiation causes transitions, transversions, deletions, and frameshift mutations in nonreplicating DNA (Drake, 1970). The mechanism of inactivation of herpesviruses by UV irradiation is thought to involve the formation of pyrimidine dimers in viral DNA (Ross et al., 1972b; Cameron, 1973). Whether the same mechanism is involved in UV-induced mutation is an open question.

Nitrous acid, which induces predominantly transitions in nonreplicating DNA through the deamination of adenine, guanine, and cytosine, has been used to obtain *ts* mutants of HSV type 2 (Zygraich and Huygelen, 1973), PRV (Huy et al., 1972), and IBR (Zygraich et al., 1974a, 1974b). Huy et al. (1972) also used 5-bromouracil, a base analogue which acts in a similar manner to BrdUrd to induce PRV *ts* mutants.

B. Isolation

The selection of a single *ts* mutant from each independently mutagenized stock eliminates the possibility of any mutants in a series being clonally related. In all but three studies with herpesvirus *ts* mutants (Timbury, 1971; Pringle et al., 1973; Manservigi, 1974), isolation of clonally related mutants was avoided in this manner.

Choosing and screening large numbers of plaque isolates to select potential *ts* mutants by classical methods is both time-consuming and laborious. Consequently, a more rapid technique for screening mutants for temperature-sensitivity would speed up the processes of genetic analysis. Robb and Martin (1970) have developed a replica-plating technique for screening potential SV40 *ts* mutants. Modifications of this technique were used by Stephenson et al. (1972) in the isolation of *ts* mutants of murine leukemia virus and by Manservigi (1974) in the isolation of HSV type 1 *ts* mutants. The replica-plating technique is highly efficient, avoids cross-contamination, and should be suitable for use with *ts* mutants of other animal viruses. Since the method relies upon transfer of virus from fluids of infected cultures to uninfected cultures the method may not be suitable for highly cell-associated viruses such as CMV and HVS.

C. Growth Properties

The difficulty in obtaining mutants with stable *ts* characteristics constitutes a serious problem in investigations with *ts* mutants. Instability of *ts* mutants is a consequence of: (a) their reversion to the ts^+ phenotype, and (b) the so-called "leaky" phenomenon in which the mutant, at the npT°, manifests the function of the WT virus. With regard to their growth properties, the herpes-virus *ts* mutants isolated to date exhibited efficiencies of plating ranging from 10^{-3} (BROWN et al., 1973) to less than 10^{-8} (TIMBURY, 1971; BROWN et al., 1973; MANSERVIGI, 1974) (Table 1). Efficiencies of replication ranged from 10^{-2} (ONODA et al., 1971) to less than 10^{-8} (MANSERVIGI, 1974). The low levels of reversion and leak exhibited by the mutants described in Table 1 indicate that most of them probably possess point mutations and are therefore suitable for genetic analysis. Mutants which do not revert and which exhibit little or no leak may, however, possess multiple mutations. The problem of reversion has necessitated care in the preparation of mutant stocks for use in genetic studies. The preparation of multiple stocks using high dilutions of previous stocks known to contain few revertants in the undiluted state (SCHAFFER et al., 1973; ESPARZA et al., 1974) or recloning from single *ts* plaques at the pT° (BROWN et al., 1973) was found to be necessary.

III. Complementation Studies

Before examining the recombination potential and phenotypic properties of a series of *ts* mutants, it is first essential to identify the minimum number of loci represented by the mutants in a series by complementation analysis. In the standard complementation test, cells are infected with two mutants and incubated at the npT° after which the yield is determined. If the *ts* mutation in each mutant is located in a different gene, the function of the defective gene product of one virus will be performed by the analogous non-defective gene product of the other. Consequently, limited virus replication will occur. If the mutants are defective in the same gene, the mixed infection will be nonproductive and the mutants are said to be in the same complementation group. Virus yields in mixed infections which are 1.5 to 2-fold higher than the sum of yields of each mutant grown separately indicate that statistically significant complementation has occurred (BURGE and PFEFFERKORN, 1966b; PADGETT and TOMKINS, 1968). By complementation analysis it has been possible to classify the series of *ts* mutants of HSV types 1 and 2 and PRV into functional groups (cistrons). Although a considerable amount of information has been obtained from genetic and phenotypic studies of *ts* mutants of these viruses, similar detailed studies with *ts* mutants of the other herpesviruses listed in Table 1 have not yet been conducted.

A. HSV Type 1

Complementation tests have been conducted with two (BROWN et al., 1973; SCHAFFER et al., 1973) of the three series of HSV type 1 *ts* mutants listed in Table 1 using one or both of the following methods: (1) An "in-fectious-center assay" which measures the ability of cells to produce infectious centers (virus-producing cells) at the npT° when simultaneously infected with

Table 2. *ts* mutants of HSV type 1

Brown et al., 1973			Schaffer et al., 1973		
Complementation group	Mutants in group	Viral DNA[a] phenotype (npT°)	Complementation group	Mutants in group[b]	Viral DNA[b] phenotype (npT°)
A	*ts*A	−	A	*ts*A1b, *ts*A15g, *ts*A16g[c]	−
B	*ts*B, *ts*B2	−, ±	B	*ts*B2b, *ts*B21u	−
C	*ts*C	+	C	*ts*C4b, *ts*C7b	−
D	*ts*D	+	D	*ts*D9b	−
F	*ts*F	+	E	*ts*E5b, *ts*E6b	+
G	*ts*G	+	F	*ts*F17g, *ts*F18g	+
I	*ts*I	+	G	*ts*G3b, *ts*G8b	+
J	*ts*J	−	H	*ts*H10b	+
			I	*ts*I11b	+
			J	*ts*J12g	+
			K	*ts*K13g	+
			L	*ts*L14g	+
			M	*ts*M19u	+
			N	*ts*N20u	+
			O	*ts*O22u	+

[a] Viral DNA phenotypes of mutants as described by Subak-Sharpe et al. (1973).

[b] Viral DNA phenotypes of mutants as described by Schaffer et al. (1973).

[c] Mutant designations: Capital letters designate complementation group; Arabic numbers designate mutant number; and lower case letters, b, g, and u designate mutagens BrdUrd, NTG, and UV-irradiation, respectively.

two different *ts* mutants (Brown et al., 1973), or (2) a "yield-of-progeny-virus test" which measures the total amount of virus produced by a population of cells mixedly infected with two different *ts* mutants at the npT°; total virus is assayed at the pT° (Brown et al., 1973; Schaffer et al., 1973).

Brown et al. (1973) obtained qualitatively similar results when complementation tests were performed by either method: the nine BrdUrd-induced *ts* mutants studied fell into eight complementation groups (Table 2). Using a multiplicity of 5 plaque-forming units (PFU) per cell of each virus in mixed infections (10 PFU/cell in single infections) in both infectious-center and yield-of-progeny-virus tests, complementation indices (CI) were calculated as follows:

$$\text{Infectious-center CI} = \frac{(ts\text{X} + ts\text{Y})^{\frac{npT°}{pT°}}}{1/2\,(ts\text{X}^{\frac{npT°}{pT°}} + ts\text{Y})^{\frac{npT°}{pT°}}} \text{ grown at npT°, where } ts\text{X}$$

and *ts*Y are any two mutants; $\frac{npT°}{pT°}$ represents the ratio of infectious- center titers at the npT° and pT°, respectively.

$$\text{Yield-of-progeny-virus CI} = \frac{(ts\text{X} + ts\text{Y})^{pT°} - (ts\text{X} + ts\text{Y})^{npT°}}{1/2\,(ts\text{X}^{pT°} + ts\text{Y})^{pT°}} \text{ grown at npT°.}$$

Yields were assayed at pT° and npT° as indicated in the formula (Brown et al., 1973).

The geometric mean of CI in infectious-center tests ranged from 0.7 for noncomplementing mutant pairs to 162 for complementing pairs and in yield-of-progeny-virus tests, from 1.1 to 85. The combination tsB plus tsE yielded CI values of 0.7 and 1.1 in infectious-center and yield-of-progeny-virus tests, respectively. These mutants were therefore placed in group B, as tsB and tsB$_2$ (formerly, tsE). Where CI in infectious-center tests were low in crosses with leaky mutants, CI values in yield-of-progeny-virus tests gave more definitive evidence of complementation. Two additional mutants, tsG and tsJ, which gave equivocal results in both types of test, were considered to complement on the following basis: The progeny of mixed infections with these two mutants, in which tsG carried the independent syn marker (tsG $syn + ts$J syn^+), yielded mixed syn, syn^+ infectious centers, indicating that both mutants were produced.

Using the yield-of-progeny-virus test, SCHAFFER et al. (1973) identified 15 non-overlapping complementation groups among 22 HSV type 1 ts mutants isolated following mutagenesis with BrdUrd, NTG and UV light (Table 2). Using a multiplicity of 2.5 PFU/cell of each virus in mixed infections (5.0 PFU/cell in single infections), CI were calculated as follows:

$$CI = \frac{(tx\text{X} + ts\text{Y})^{p\text{T}°}}{(ts\text{X}^{p\text{T}°} + ts\text{Y})^{p\text{T}°}} ,$$

where tsX and tsY were any two mutants; infected cultures were incubated at npT° and assayed at pT°. CI values of 2 or greater were considered to be positive (SCHAFFER et al., 1973).

Complementation indices ranged from 0.5 (noncomplementing pairs) to 9300 (complementing pairs). Generally speaking, all mutants belonging to groups with more than one member complemented other mutants to approximately the same degree. Complementation yields [CY $= (ts\text{X}+ts\text{Y})^{np\text{T}°} - (ts\text{X}^{np\text{T}°}+ts\text{Y}^{np\text{T}°})$ assayed at pT°] ranged from 1–5% of the yield of WT virus grown under identical conditions. With efficiently complementing pairs this value occasionally reached 10%. It is interesting to note that the three agents used for mutagenesis were able to induce ts lesions within the same viral gene in two cases (groups A and B) (Table 2).

The results of complementation analyses of HSV type 1 ts mutants have demonstrated several noteworthy points regarding the quality of the tests and mutants employed, and the genetic composition of the HSV type 1 genome: (1) The yield-of-progeny-virus test was superior to the infectious-center test in both the magnitude of the complementation indices obtained and in the clarity of complementation demonstrated. Furthermore, a potential problem of the infectious-center test is that if recombination could occur in the absence of complementation, and recombinants give rise to infectious centers at the npT°, the test would measure recombinant-initiated infectious centers. If this were so, the test would yield false-positive results (BROWN et al., 1973). In addition, the yield-of-progeny-virus test would be expected to give more definitive proof of complementation than the infectious-center test if there were a significant proportion of recombinant-initiated centers because the relative contribution of ts^+ recombinants in the final complementation yield would be reduced. Brown et al. (1973) have suggested, however, that the frequency of recombinant-initiated infectious centers was negligible in their study. (2) Some variability in complementation indices from test to test was noted in both studies; however, indices in yield-of-progeny-virus tests were significantly higher, and less variable in repeat tests in the studies of Schaffer et al. (1973). (3) The fact that non-overlapping patterns of complementation were observed with both series of mutants suggests that the mutants in these studies possessed single defects. (4) If we assume that some, but not all, of the complementation groups from each laboratory will eventually prove to be identical, the number of HSV type 1 cistrons for which mutants have been isolated is greater than 15. With a coding capacity of about 80 polypeptides, the identification of over 15 genes represents more than 20% of the coding capacity of the HSV type 1 genome. (5) Brown et al. (1973), commenting on the fact that 8 out of the first 9 of their mutants fell into different complementation groups, suggested on the basis of the Poisson distribution that the HSV genome contains over 30 cistrons with essential functions at the npT°.

B. HSV Type 2

Of the four groups of HSV type 2 ts mutants described so far (Table 1), only two (Timbury, 1971; Esparza et al., 1974) have been subjected to complementation analysis. Of the 33 BrdUrd-induced mutants isolated by Timbury (1971), 10 have been examined in detail and assigned to 10 complementation groups using an infectious-center test similar to that described for ts mutants of HSV type 1 (Brown et al., 1973) (Table 3). Although tests for complementation were first attempted by comparing yields of virus from singly and mixedly infected monolayers (yield of progeny virus test, Brown et al., 1973), no complementation could be demonstrated by this method. Using the infectious-center test, the degree of complementation varied considerably in different experiments and negative results were occasionally obtained with pairs of mutants which had previously demonstrated significant complementation. The author states that 12 additional mutants belong to the 10 groups previously identified. This would not be surprising since the WT virus was not reported to have been cloned before mutagenesis and all mutants were selected from a single mutagenized stock. Conditions, therefore, favored the isolation of multiple pre-existing and/or "sister" ts mutants (Timbury, 1971). Although the infectious-center test proved to be less reliable for classifying ts mutants of HSV type 1 than the yield of progeny virus test (Brown et al., 1973), without this method the cistron assignments of this series of HSV type 2 ts mutants could not have been made.

A second series of 8 BrdUrd-induced HSV type 2 ts mutants was unambiguously assigned to 7 non-overlapping complementation groups in a yield of progeny virus test by Esparza et al. (1974) (Table 3). In this case

Table 3. *ts* mutants of HSV type 2

TIMBURY, 1971			ESPARZA et al., 1973		
Comple-mentation group	Mutants in group	Viral DNA[a] phenotype (npT°)	Comple-mentation group	Mutants in group[b]	Viral DNA[c] phenotype (npT°)
1	*ts*1	—	A	*ts*A1b, *ts*A8b	—
2	*ts*2	—	B	*ts*B5b	—
3	*ts*3	+	C	*ts*C2b	—
4	*ts*4	+	D	*ts*D6b	+
5	*ts*5	+	E	*ts*E7b	+
6	*ts*6	—	F	*ts*F3b	+
7	*ts*7	—	G	*ts*G4b	+
8	*ts*8	—			
9	*ts*9	—			
10	*ts*10	—			

[a] Viral DNA phenotypes as described by HALLIBURTON and TIMBURY (1973).
[b] Mutant designations are as described in footnote "c" of Table 2 for HSV type 1 *ts* mutants.
[c] Viral DNA phenotypes as described by ESPARZA et al. (1973).

efficient complementation was demonstrated with most mutant pairs. Complementation indices ranged from 0.3 for noncomplementing pairs to 6000 for complementing pairs. As stated by the authors, the differences observed in the magnitude of complementation obtained in their study and in that of TIMBURY (1971) could be due to differences in cells, virus strains, or methodology used in the two investigations, or to the ultimate nature of the defects of the various mutants. Complementation yields [CY $= (tsX+tsY)^{npT°} - (tsX^{npT°}+tsY^{npT°})$ assayed at pT°] among complementing mutant pairs ranged from $<1\%$ to 10% of the WT virus yield obtained under identical conditions.

C. PRV

PRINGLE et al. (1973) have classified 10 BrdUrd-induced *ts* mutants of PRV (Table 1) into 9 complementation groups using a yield of progeny virus test which had been developed for vesicular stomatitis virus (PRINGLE, 1970). Complementation indices were calculated using the formula described by SCHAFFER et al. (1973) (see Section *III. A.*, above). Complementation was highly efficient in that indices ranging from 1 to 560000 were obtained.

HUY et al. (1972) (Table 1) have described 9 nitrous acid- and bromouracil-induced *ts* mutants of PRV isolated using a selective technique involving temperature down-shift. Since only 4 complementation groups were identified among the 9 mutants, it is possible that the isolation procedure may have been more restrictive than isolation without selection by favoring mutants defective late in the infectious cycle (PRINGLE et al., 1973).

The available data concerning complementation with *ts* mutants of HSV type 1, type 2, and PRV have shown that extensive exchange of gene products occurs between *ts* mutant pairs as demonstrated by the high complementation indices and yields obtained.

For the complete functional analysis of a virus it is necessary to obtain and identify mutants which are defective in all essential cistrons. The large size of the herpesvirus genome clearly makes such isolation and complementation attempts a formidable endeavor. We may assume that to date less than 50% of the essential gene functions of HSV type 1, and an even smaller percentage of the essential functions of HSV type 2 and PRV have been identified by means of quantitative complementation tests. It is not known whether additional cistrons are represented by the existing *ts* mutants of HSV type 1, type 2, and PRV since complementation tests between mutants from different laboratories have not been performed. Efforts to identify new mutants of these viruses must include complementation tests with members of all existing complementation groups. While quantitative complementation assays are definitive in assigning mutants to complementation groups, they are laborious. The task of identifying herpesvirus cistrons in new and existing mutants would be greatly facilitated by the availability of a qualitative complementation test which would be applicable over a wide range of virus titers and which would accommodate mutants with low and moderate leak or reversion frequencies.

Recently two qualitative complementation tests have been described which meet these criteria and which have been used successfully to classify *ts* mutants of SV40 (Chou and Martin, 1974; Tevethia et al., 1974). Both tests depend upon the ability of the mutant to diffuse through agar. Since SV40 is both smaller (more freely diffusable) and more thermostable than the herpesviruses, the feasibility of using these tests with the herpesviruses remains to be seen.

IV. Recombination Studies and Linkage Maps

The ability of *ts* mutants to interact by recombination offers two distinct advantages for their genetic analysis. One advantage is the confirmation by recombination of cistron assignments made by complementation analysis since mutants defective in the same cistron (noncomplementing mutants) should not be able to generate *ts*+ recombinants efficiently. The second advantage is the ability to order cistrons through the construction of genetic maps which can then yield information concerning the organization and regulation of virus functions. It has been known for some time that recombination occurs between strains of HSV (Wildy, 1955). Recombination analysis and the construction of provisional linkage maps have been carried out with *ts* mutants of HSV type 1, type 2, and PRV.

A. HSV Type 1

In the study of Brown et al. (1973) recombination analysis was conducted using a technique involving 3-factor crosses. As mentioned previously, each *ts* mutant in this series was available in both the *ts* X *syn* and *ts* X *syn*+ forms.

Therefore, in addition to the selected *ts* markers in the cross, the non-selected *syn* marker was utilized. The experimental procedure was that described above by BROWN et al. (1973) for the yield of progeny virus test in complementation studies with two exceptions: a) infected cultures were incubated at the pT°, and b) in each test, three input multiplicities of each mutant were utilized so as to yield a *syn/syn*⁺ ratio in the progeny of no greater than 3:1 nor less than 1:3. The latter procedure was introduced to minimize nonstandard recombination frequencies which can arise from deviations in the input multiplicity of genetically active genomes (VISCONTI and DELBRÜCK, 1953). Furthermore, since recombination between the genomes of noninfectious particles may lead to the production of *ts*⁺ recombinants, BROWN et al. (1973) used mutant stocks with particle/PFU ratios as close to unity as possible. In this way the contribution of the genomes of noninfectious particles was minimized since the input multiplicity was calculated on the basis of infectious units (PFU). The progeny of recombination tests were assayed at the pT° (to measure total virus) and npT° (to measure *ts*⁺ recombinants). Recombination frequencies (RF) were calculated from the relationship.

$$\text{RF } (\%) = 100 \times 2 \left[(ts\text{X}syn \times ts\text{Y}syn^+)^{\frac{npT°}{pT°}} - 1/2 \left(ts\text{X}syn^{\frac{npT°}{pT°}} + ts\text{Y}syn^{+\frac{npT°}{pT°}} \right) \right]$$

where *ts*Xsyn and *ts*Ysyn⁺ represent any pair of mutants; one in the *syn* and one in the *syn*⁺ form. Recombinants were scored as *syn* or *syn*⁺ to permit ordering of the *syn* gene in relation to the two parental *ts* markers. When 9 mutants were crossed, all but two, *ts*B and *ts*B2, which had been assigned to the same complementation group, yielded significant numbers of *ts*⁺ recombinants. The proportion of recombinants produced in all crosses ranged from less than 0.0025% (RF = <0.005%, *ts*B × *ts*B2), to 14.5% (RF = 29%, *ts*F × *ts*D). Recombination data thus confirmed the assignment of the 9 mutants to 8 complementation groups. The published results from this study represent the data obtained from a single, simultaneous test which included all possible crosses (BROWN et al., 1973). Relative recombination frequencies from different tests, however, remained fairly constant. Assays of putative *ts*⁺ recombinant plaque isolates in progeny at both the pT° and npT° demonstrated that the great majority were true recombinants, i.e. the ratio of plaques produced at npT°/pT° approached unity. In progeny tests the authors noted the presence of *ts*⁺ plaques with mixed *syn*, ·*syn*⁺ morphology. Although most mixed plaques segregated upon replication to produce true-breeding *syn* or *syn*⁺ plaques, some yielded mixed plaques. These findings suggest the possibility that the particles which generated mixed plaques were at least partially heterozygous and arose as a consequence of recombination (SUBAK-SHARPE, 1973). As the authors state, however, the percent recombinants varied as much as 10-fold from test to test, depending upon: a) the "effective genome" input, and b) the number of rounds of mating which had occurred. The latter variable was apparently dependent upon the physiological state of the cells and the time of incubation. The greatest rate increase in the generation of recombinants occurred from 6 to 10 hours post-infection.

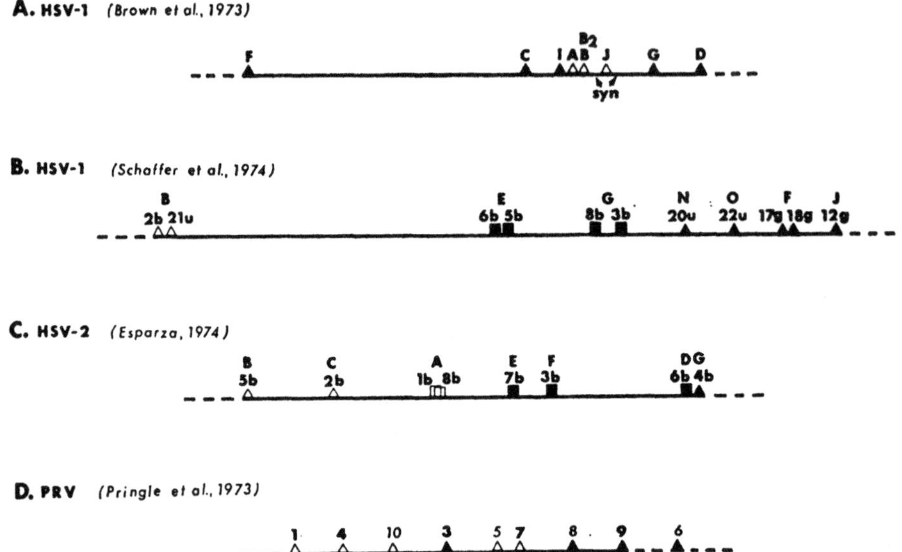

Fig. 1 A–D. Linkage maps of HSV types 1 and 2 and PRV. (A) Linkage map of 9 *ts* mutants of HSV type 1 including the *syn* marker. The map spans about 25 units (BROWN et al., 1973). (B) Linkage map of 11 *ts* mutants of HSV type 1. The map spans approximately 38 units (SCHAFFER et al., 1974a). (C) Linkage map of 8 *ts* mutants of HSV type 2 spanning approximately 25 units (ESPARZA, 1974). (D) Provisional linkage map of 9 *ts* mutants of PRV showing gene order only. The total map length and distances between markers are considered to be tentative by the author and were not included in the figure (PRINGLE et al., 1973). The maps in A, B, and C are drawn to scale. *Syn* syncytial marker; △ DNA⁻ mutants; ▲ DNA⁺ mutants; □ DNA⁻ mutants with thermosensitive viral structural components; ■ DNA⁺ mutants with thermosensitive viral structural components; Mutant designations: Capital letters = cistrons; Arabic numbers = mutant numbers (numbers in italics in D = mutants whose order is provisional); Small letters = agents used for mutant induction: b = BrdUrd, g = NTG, and U = UV-irradiation

The recombination frequencies obtained from 3-factor crosses have permitted the ordering of 9 *ts* mutants (8 cistrons) on a provisional linkage map (Fig. 1A). The linkage map so obtained spans approximately 25 map units; 7 of the 8 cistrons fell within a distance of 9 units (sum of the shortest intervals), and certain intercistronic distances were found to be less than 1 unit. The use of 3-factor crosses in this study thus reflects a unique advantage of such crosses, i.e. the reliable ordering of cistrons over short intervals. The map so far shows no evidence of circularity and mutants in the 3 DNA⁻ complementation groups appear to cluster in the middle of the map within a span of 3 map units (BROWN et al., 1973; SUBAK-SHARPE, 1973).

With a second series of HSV type 1 *ts* mutants, SCHAFFER et al. (1974) examined the recombination potential of 15 *ts* mutants representing 10 complementation groups using 2-factor crosses. The experimental procedure for these studies was the same as that described for complementation with these mutants (SCHAFFER et al., 1973), except that infected cultures were incubated at the pT°. Mutant stocks with particle/PFU ratios as close to unity as possible were used in order to minimize the contribution of noninfectious genomes to the input inoculum and inocula were assayed immediately after infection of cultures to confirm calculated input multiplicities. Recombination frequencies (RF) were calculated using the following formula:

$$RF\ (\%) = 100 \times 2\ \frac{(ts\mathrm{X} \times ts\mathrm{Y})^{\mathrm{npT}^\circ}}{(ts\mathrm{X} \times ts\mathrm{Y})^{\mathrm{pT}^\circ}}$$

where mixedly infected cultures were incubated at pT° and assayed at pT° and npT° as indicated in the formula. The RF obtained were presented as the average of from 3 to 5 separate determinations of each cross. Recombination frequencies ranging from less than 1% to greater than 50% were reproducibly found. Mutants in the same complementation group either failed to recombine or recombined with low frequency. Furthermore, mutants in the same complementation group recombined with similar frequencies with other mutants. Thus, recombination data confirmed the assignment of mutants to cistrons by complementation analysis. Progeny of mixed infections were tested for the possibility that complementing clumps of the two ts mutants in a given cross could give rise to plaques at the npT°. Mixed yields were therefore subjected to sonication (to disrupt clumps of ts mutants) and filtration (to remove clumps). Neither treatment alone nor both in combination demonstrated the existence of significant numbers of mixed clumps. However, since minimal clumping did occur, yields from all mixed infections were routinely sonicated before assay. Putative ts^+ recombinant plaques were picked and plated at pT° and npT°. As in the studies of Brown et al. (1973), the great majority of plaques behaved as ts^+ recombinants and not as plaques which arose from complementing clumps of both ts mutants or multiploid particles containing the genomes of both mutants.

A provisional linkage map was constructed using the average RF obtained in 3 to 5 crosses involving mutants in 7 of the 10 complementation groups studied (Fig. 1 B). A linear map, spanning approximately 38 units, was obtained. Six of the 7 cistrons fell within a span of 20 units; the seventh cistron was located 20 units from its nearest neighbor. Mutants in DNA⁻ complementation groups were located at the left-hand end of the map and mutants in two groups, which apparently synthesize temperature-sensitive structural polypeptides (see Section $VI. F$, below), fell in the center of the map. DNA⁺, thermostable mutants were located at the right-hand end of the map. These mutants thus appear to be defective in some late step in replication involving encapsidation, envelopment, or some other step in the maturation of virus particles.

B. HSV Type 2

Esparza (1974) obtained efficient recombination in 2-factor crosses between 8 previously described ts mutants Esparza et al. (1974) using the method described by Schaffer et al. (1974). The 8 mutants demonstrated recombination frequencies ranging from less than 1% to about 20%. The two mutants which failed to complement (Esparza et al., 1974) recombined with the lowest frequency.

A provisional, linear linkage map containing 7 cistrons (8 mutants) and spanning 25 map units (sum of the shortest intervals) has been constructed (Fig. 1C). DNA$^-$ mutants were located at the left-hand end of the map and those exhibiting thermosensitive virion structural polypeptides, at the right-hand end. The only mutant which resembled the WT virus phenotypically was located at the right-hand end of the map and probably possessed a very late defect. Thus, clustering of functionally related genes was also a property of this linkage map.

C. PRV

Efficient recombination between PRV *ts* mutants in 2-factor crosses was demonstrated by PRINGLE et al. (1973). The experimental procedure was identical to that used in complementation studies with these mutants (PRINGLE et al., 1973) except that incubation was carried out at the pT°. RF (%) were calculated from the relation:

$$\mathrm{RF}\ \% = 100 \times 2 \times \left\{ \frac{(ts\mathrm{X} \times ts\mathrm{Y})^{\mathrm{npT°}}}{(ts\mathrm{X} \times ts\mathrm{Y})^{\mathrm{pT°}}} \right\}$$

where X and Y were any pair of *ts* mutants. Values were presented as the mean values obtained in 2 or 3 separate experiments. The percentage of *ts*$^+$ recombinants increased sharply between 6 and 8 hours post-infection and stabilized after 12 hours. Mean RF ranged from 0.03% to 43% and most mutant pairs exhibited appreciable recombination. Mutants *ts*5 and *ts*7 which belong to the same complementation group recombined inefficiently (0.08%). Upon recloning, all putative *ts*$^+$ recombinants behaved phenotypically as WT virus. Using selected data from all possible crosses, 9 mutants were tentatively ordered in linear array (Fig. 1D). Although the recombination frequencies obtained demonstrated some additivity, the additivity was insufficient to allow relative distances to be indicated on the map. The authors emphasized the provisional nature of both the order of loci on the map and the distances between them. It is of interest to note, however, that the 5 mutants defective in viral DNA synthesis were located at the left-hand end of the map.

The results of recombination studies and mapping attempts with *ts* mutants of HSV type 1, type 2, and PRV have clearly demonstrated the following: (1) Efficient recombination occurs between most *ts* mutant pairs. (2) Mutants defective in the same cistron do not recombine efficiently; however, in crosses with other mutants they recombine with similar frequencies. (3) Recombination frequencies were sufficiently additive to permit the ordering of cistrons to form provisional linkage maps.

A comparison of results obtained by 2- and 3-factor crosses has demonstrated that either method can be used to order cistrons for the construction of genetic maps. The method of 3-factor crosses offers the distinct advantage that both the order of markers and the quantification of map distances can be determined with greater confidence than with 2-factor crosses. Hence, in cases where mutants with the required properties are available, 3-factor crosses should be attempted. It should be noted, however, that when the same series of HSV type 1 *ts* mutants was analyzed by both kinds of test, the sequential order of genes was affected in only one instance and although the absolute frequencies of recombination were different, relative frequencies were the same (HAY et al., 1971; BROWN et al., 1973). Furthermore, studies utilizing 2-factor crosses have provided valuable information regarding genome organization and mechanisms of recombination in both bacterial virus systems (HAYES, 1968) and more recently in animal virus systems (FENNER, 1970).

The investigators who formulated the genetic maps described above, all stressed the provisional nature of the maps. Consequently, the comparisons presented in Fig. 1 should be interpreted with great caution. However, the clustering of functionally related genes in the same general pattern appears to be more than fortuitous. DNA⁻ cistrons are located to the left in three of the four maps. The clustering of DNA⁻ cistrons in the center of the map of BROWN et al. (1973) could indicate that these authors have isolated additional mutants which fall to the left of the region responsible for the DNA⁻ phenotype. The clustering of DNA⁻ cistrons on the maps of HSV types 1 and 2 is not surprising in view of their presumed close phylogenetic relationship. It is of interest that the map of a more distantly related virus, PRV, has also demonstrated similar clustering of DNA⁻ cistrons (PRINGLE et al., 1973).

On the HSV type 1 linkage map of SCHAFFER et al. (1974), all *ts* mutants induced with NTG (Table 2, *ts*J12, *ts*F17, and *ts*F18) and 2 of 3 mutants induced with UV irradiation (*ts*N20 and *ts*O22) are DNA⁺ and lie on the right-hand end of the map (Fig. 1B). This observation suggests that mutagenesis of HSV type 1 with NTG and UV light may result in the selective induction of *ts* mutants defective in functions expressed *after* viral DNA synthesis, e.g. maturation. Experiments are in progress with NTG- and UV-induced *ts* mutants of HSV type 2, most of which are DNA⁺, to determine whether these mutants will map in a pattern similar to the HSV type 1 mutants. In addition, since most mutants included in mapping experiments were BrdUrd-induced (with the exception of the UV- and NTG-induced mutants of SCHAFFER et al., 1974), and since BrdUrd exhibits greater "hot-spotting" than other mutagens, the phenotypic properties and the map location of mutants may reflect the mutagenic specificity of BrdUrd.

V. Complementation and Recombination Between *ts* Mutants of HSV-1 and HSV-2

The classification of herpes simplex viruses into two types was based initially on antigenic differences observed among isolates of HSV as demonstrated by neutralization tests (NAHMIAS and DOWDLE, 1968). Later studies demonstrated that virus strains representative of the two serotypes also differed in a number of biological, biophysical, and biochemical characteristics (FIGUEORA and RAWLS, 1969; ROIZMAN et al., 1972). HSV type 2 has been associated with cervical carcinoma (RAWLS et al., 1968; NAHMIAS et al., 1970; AURELIAN, 1973), and both HSV types 1 and 2 can induce oncogenic transformation of hamster cells *in vitro* (DUFF and RAPP, 1971, 1973). Nucleic-acid hybridization studies have provided a molecular basis for the phenotypic differences observed between strains of HSV types 1 and 2. As previously mentioned, the DNAs of these viruses share approximately 50% of their nucleotide sequences in common (KIEFF et al., 1972; LUDWIG et al., 1972a). Furthermore, KIEFF et al. (1972) observed that there are two classes of DNA sequences in HSV-DNA. One class that comprises 53% of the DNA apparently consists of predominantly type-specific sequences. The second class, comprising 47% of the DNA consists of homologous regions with good matching of base pairs (85%). The evolutionary significance of variable and invariant regions of the two DNAs is still unknown. Studies of the genetic interactions between HSV type 1 and 2 *ts* mutants, however, offer a unique opportunity: a) to a better understanding of the degree of functional homology which exists between these viruses, and b) to a comparison of the physical organization of the two genomes.

Intertypic complementation tests were performed to examine whether, under npT° conditions, *ts* mutants of HSV type 1 and 2 could provide reciprocally the normal functional gene product(s) necessary for *ts* mutant replication. Intertypic recombination experiments were conducted: a) to determine whether HSV type 1–HSV type 2 *ts* mutant pairs could generate stable *ts*+ virus efficiently, and b) to study the absolute and relative recombination frequencies obtained in intertypic recombination experiments as compared with those obtained for HSV type 1 and type 2 in *intra*typic recombination.

The interaction of HSV type 1 and type 2 *ts* mutants in complementation and recombination tests has been reported by TIMBURY and SUBAK-SHARPE (1973), ESPARZA et al. (1973), and BENYESH-MELNICK et al. (personal communication). TIMBURY and SUBAK-SHARPE (1973) described intertypic complementation experiments using 7 *ts* mutants of HSV type 1 and 10 *ts* mutants of HSV type 2 using an infectious-center test. All HSV type 1 mutants in these tests carried the *syn* marker. Complementation was efficient when it occurred; however, in a few instances negative results were obtained with mutants which had previously complemented well. All but 4 mutants failed to complement at least one, and usually two or more mutants of the opposite type. Some mutant pairs which failed to complement possessed different DNA phenotypes at the npT°. Since the lack of complementation was observed in an unexpectedly large number of cases, it was suggested that failure to complement was not likely to be due simply to the *ts* defects being in corresponding cistrons of the two noncomplementing mutant genomes, but rather, perhaps, to complex interactions involving genes with regulatory functions. The authors, however, failed to suggest the possibility that at least some of the observed lack of complementation could have been due to mutants of either or both types which possessed multiple *ts* defects. Analysis of the progeny within infectious centers produced by two complementing mutant pairs at the npT° was carried out with the following results.

a) Both parental mutants (HSV type 1 *ts*X*syn* and HSV type 2 *ts*Y*syn*⁺) were recovered indicating that both genomes had replicated by complementation mechanisms, b) in a few cases, presumptive *ts*⁺ recombinants were recovered, some of which continued to segregate with regard to the *syn* and/or *ts* markers for at least three progeny generations, and c) a number of these clones exhibited phenotypes intermediate between HSV type 1 and type 2 with respect to thermal stability, plaque morphology, and differential neutralization of infectivity.

In a second study of intertypic complementation and recombination, 7 mutants of HSV type 1 (representing 5 complementation groups) and 7 mutants of HSV type 2 (representing 7 complementation groups) were employed by ESPARZA (1974) (Table 4). Mutants defective in two of the HSV type 1 cistrons and three of the HSV type 2 cistrons were unable to synthesize viral DNA at the npT°. Two HSV type 1 complementation groups, one DNA⁻ and one DNA⁺ (C and E, respectively) were each composed of two *ts* mutants. These pairs were included as internal controls, with the assumption that if intertypic complementation and/or recombination were observed, it should occur with both mutants within a group at about the same level.

The procedures used for intertypic complementation were identical to those used for *intra*typic complementation (the yield-of-progeny-virus test) (ESPARZA et al., 1974), the only difference being that intertypic complementation was carried out at 38° (the npT° for HSV type 2 *ts* mutants) because 39° (the npT° for HSV type 1 *ts* mutants) significantly inhibited the growth of the HSV type 2 WT virus. Assays were performed at the pT°. Efficient complementation was detected between most mutant pairs (Table 4). Between many complementing mutants complementation was as efficient as in intratypic tests; CI values ranged from 2.5 to > 16000. The CI obtained in mixed infections containing HSV type 1 DNA⁺ mutants were consistently lower than those obtained in mixed infections containing HSV type 1 DNA⁻ mutants. In calculating the CI for each mixed infection, this was found to be due to the occurrence of increased levels of leakiness in single, control DNA⁺ mutant infections at 38°. In contrast only low levels of leak were observed with the DNA⁻ mutants at this temperature. As expected, the CI obtained with HSV type 1 mutants belonging to the same complementation groups (C and E) were roughly similar.

Mutants in 3 of 5 HSV type 1 complementation groups failed to complement mutants in 3 of 7 HSV type 2 complementation groups (Table 4, CI values in boxes). As in the studies of TIMBURY and SUBAK-SHARPE (1973), the number of noncomplementing pairs was unexpectedly large (5 out of 49 mutant pairs or 3 out of 35 possible cistron pairs). The phenotype of the HSV type 1 + type 2 noncomplementing mutant pairs was similar with regard to DNA synthesis at the npT°. However, noncomplementing mutants were

Table 4. *Intertypic complementation and recombination between* ts *mutants of HSV type 1 and type 2*

HSV type 1		HSV type 2						
		A	B	C	D	E	F	G
		*ts*8b	*ts*5b	*ts*2b	*ts*6B	*ts*7b	*ts*3b	*ts*4b
	DNA	−	−	−	+	+	+	+
B *ts*2B	−	183[a] (1.5)[b]	576 (6.6)	166 (1.5)	130 (1.8)	146 (0.9)	380 (3.5)	273 (3.1)
C *ts*4b	−	1053	0.6 (0.2)	1250	233	7701	16832	6429
*ts*7b	−	4166	0.5 (0.005)	849	8710	607	1119	6176
E *ts*5b	+	30 (4.2)	39 (9.4)	30 (4.9)	2.5 (1.5)	7.5 (3.2)	75 (5.3)	0.5 (0.1)
*ts*6b	+	16	47	16	3	7.9	71	0.3 (0.4)
F *ts*18g	+	8.6 (3.1)	13 (1.4)	9.5 (1.2)	6.2 (1.2)	7.6 (0.6)	36 (1.8)	26 (2.8)
G *ts*3b	+	7.5	19	3.9	0.4 (<0.003)	43	25	18

[a] Complementation indices were determined by a yield of progeny virus test conducted at 38°, the npT° for HSV type 2 ts mutants, as described by Esparza et al. (1973):

$$CI = \frac{(tsX + tsY)^{npT°}}{(tsX)^{npT°} + (tsY)^{npT°}} \text{ assayed at pT°.}$$

[b] Recombination frequencies (values in parentheses) were determined as described by Schaffer et al. (1974):

$$RF = 100 \times 2 \times \frac{[(tsX \times tsY)^{npT°}}{(tsX \times tsY)^{pT°}} \text{ grown at pT° and assayed as indicated in the formula.}$$

phenotypically different with regard to several other parameters (see Table 6, *Section VI*, below).

The inability of mutants to complement suggested that functionally equivalent genes were defective in both mutants. However, this classical interpretation cannot be applied directly to the intertypic system. A simplified model of the possible mechanisms which could play a role in intertypic complementation tests is shown in Fig. 2. The definition of the hypothetical "equivalent" or "common" and "type-specific" genes to be discussed are functional: A "common" gene is one in which the DNAs of both HSV type 1 and type 2 share sufficient base sequence homology to code for proteins which can perform the same function interchangeably. Common gene products encoded by either HSV type 1 or type 2 would not necessarily have identical amino-acid sequences nor the same antigenic characteristics. Indeed, evidence has been presented that several HSV-induced proteins, including HSV-induced thymidine kinase (Thouless, 1972), and most HSV-induced glyco-

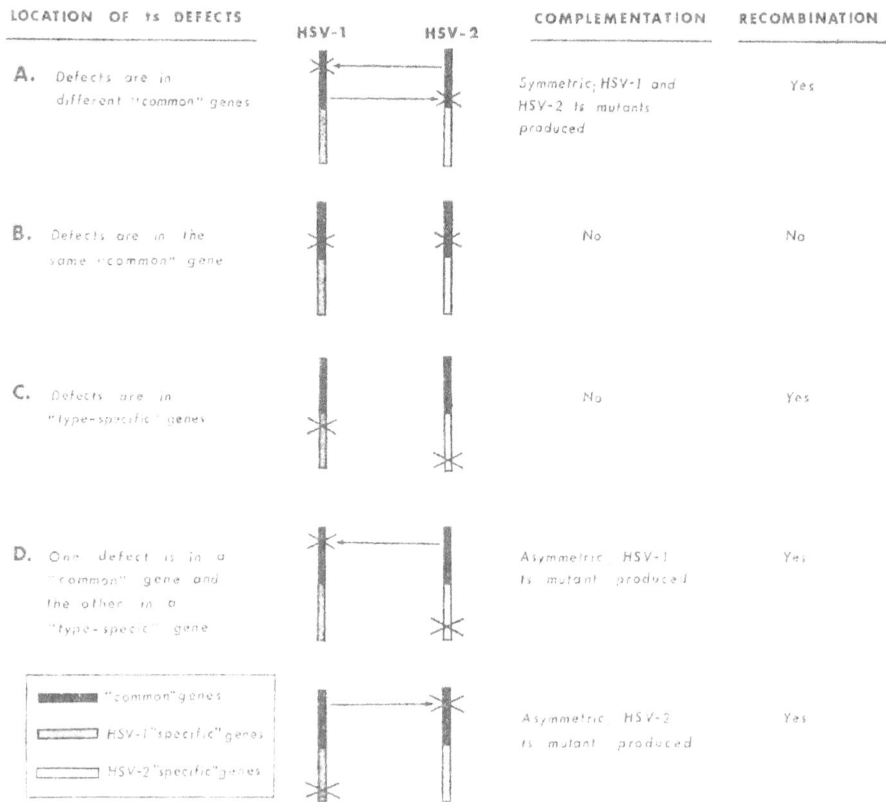

Fig. 2. A simplified model of the possible mechanisms in intertypic complementation and consequences for intertypic recombination

proteins (SAVAGE et al., 1972), including the "Band II protein" (SIM and WATSON, 1973), may possess both common and type-specific antigenic determinants. "Type-specific" genes, on the other hand, are not functionally interchangeable. The degree of "equivalence" or "type-specificity" probably varies considerably from gene to gene.

If in an HSV type 1 + type 2 *ts* mutant pair, defects in both mutants were located in different common genes, symmetric complementation would occur, with both HSV type 1 and type 2 *ts* mutants being produced (Fig. 2A). Furthermore, *ts*+ virus could be formed between these mutants by recombination events at the sites of genetic homology in the HSV type 1 and type 2 mutant genomes. If both defects were located in analogous common genes, no complementation would occur, and *ts*+ virus could not be formed efficiently through recombination (Fig. 2B).

Non-classical behavior in intertypic complementation tests would be observed if mutations occurred in hypothetical, type-specific genes (Fig. 2C). If the defects in both mutants were located in type-specific genes, no complementation would occur because the mutants involved in the cross could not provide for each other the gene products needed for normal replication. However, since the defects are in different genes, recombination could occur in homologous regions and ts^+ virus could be generated.

The fourth possibility to be considered is that in which one ts defect is located in a common gene and the other in a type-specific gene. In this case, asymmetric complementation should occur with the production of only that mutant which was defective in the common gene (Fig. 2D). Thus, the lack of complementation observed in the three cases shown in Table 4 could be explained by the situations shown in Figs. 2B and 2C.

As shown in Table 4, all heterologous mutants which failed to complement each other were able to complement other mutants with high efficiency. Two possibilities could account for this observation: a) the two mutants in a complementing pair were defective in different common genes, in which case symmetric complementation would occur with both HSV type 1 and type 2 mutants being produced (Fig. 2A), or b) one mutant was defective in a common gene and the other in a type-specific gene, in which case asymmetric complementation would occur with the production of only that mutant which was defective in the common gene (Fig. 2D).

In order to determine whether the progeny of complementing mutant pairs was symmetric or asymmetric, the temperature sensitivity and antigenic serotype of progeny virus from complementing heterotypic mutant pairs were examined. Each complementing mixed infection involved at least one mutant which had failed to complement other mutants in the intertypic test (HSV type 1 tsC7, tsE5 and tsG3; HSV type 2 tsB5, tsD6 and tsG4, Tables 4 and 5). To characterize the temperature sensitivity and antigenic phenotype of progeny virus produced in these mixed infections, individual plaque isolates from complementation yields were examined for their ability to replicate at the $pT°$ and $npT°$. Temperature-sensitive isolates were analyzed by neutralization using HSV type 1 and type 2 antisera rendered type-specfic by cross-absorption with cells infected by the heterologous WT virus (ESPARZA et al., in preparation).

Complementation yields were symmetrical in 8 out of 9 cases tested, since roughly the same proportion of mutant progeny with the antigenic phenotype of HSV type 1 and type 2 was present (Table 5). It should be noted that plaques were isolated from terminal dilutions of progeny yields (i.e., 10^{-5} to 10^{-6}) and that the proportion of phenotypically HSV type 1 and type 2 ts virions at these dilutions reflected the proportion in the total population. The data obtained from progeny analysis were thus compatible with the concept that noncomplementing HSV type 1 + type 2 mutant pairs were defective in common or equivalent genes (Fig. 2A), and that no mutants defective in strictly type-specific functions were identified in intertypic complementation tests. The reason for the apparent asymmetry in one case (HSV type 1 tsC7 + HSV type 2 tsG4) is presently being investigated. In addition, complementation studies conducted between HSV type 1 and type 2 WT viruses and mutants of the heterologous type have indicated that the WT

Table 5. Progeny analysis from intertypic complementation tests

ts mutant		Ratio HSV type 1/type 2
HSV-1	HSV-2	*ts* plaque isolates[a]
*ts*7b	*ts*5	N.C.[b]
*ts*7b	*ts*8b	11/16
*ts*7b	*ts*2b	12/9
*ts*7b	*ts*6b	14/6
*ts*7b	*ts*4b	0/17
*ts*5b	*ts*4b	N.C.
*ts*5b	*ts*5b	13/14
*ts*5b	*ts*2b	17/6
*ts*5b	*ts*6b	18/5
*ts*3b	*ts*6b	N.C.
*ts*3b	*ts*2b	9/14
*ts*3b	*ts*4b	5/16

[a] *Ts* plaque isolates were tested to determine which were antigenically HSV type 1 and which were type 2. *Ts*+ recombinant plaque isolates were not tested for antigenic type. Values are presented as the ratio of *ts* isolates which were antigenically HSV type 1 to those which were antigenically HSV type 2.
[b] N.C. = noncomplementing mutants, no progeny analysis.

virus was able to complement most heterotypic *ts* mutants and antigenic analysis of progeny also failed to reveal the existence of type-specific mutations (ESPARZA, 1974).

Although recombination analysis offers a reliable way to determine the physical distances between markers when studies are conducted with mutants of the same organism, the situation becomes more complex when only partially homologous DNAs interact in recombination. Recombination frequencies of partially homologous DNAs will be influenced by the physical distances between markers and by the degree of homology, i.e. the total number of homologous base sequences and the length, number, and distribution of homologous regions.

Selected heterologous HSV type 1 and type 2 mutant pairs used in complementation tests were subjected to recombination analysis (ESPARZA, 1974) (Table 4, values in parentheses). The intertypic recombination frequencies obtained were consistently lower than those observed in intratypic tests (ESPARZA, 1974; SCHAFFER et al., 1974), and failed to demonstrate sufficient additivity for the ordering of markers. Both of these observations may reflect the limited genetic homology which exists between the DNAs of HSV types 1 and 2. As predicted by the model in Fig. 2B and as indicated from intertypic complementation experiments, noncomplementing HSV type 1 and type 2 *ts* mutant pairs were unable to generate significant numbers of *ts*+ virus through recombination.

The above studies thus failed to demonstrate the existence of solely type-specific *ts* mutations which may be due either to the relative scarcity of these genes or to the fact that only common gene-functions are essential for HSV replication.

The intertypic complementation studies of Timbury and Subak-Sharpe (1973) and of Esparza (1974) have demonstrated that the DNAs of HSV type 1 and type 2 share sufficient base-sequence homology to code for functionally interchangeable gene products. However, failure of an unexpectedly large number of heterologous mutant pairs to complement and the significant differences in the phenotypes of mutants in noncomplementing pairs in both of these studies suggests basic differences in the regulatory mechanisms of the two viruses.

Although stable *ts*⁺ recombinants were generated by intertypic recombination, the frequency of recombination was lower than in intratypic tests and the frequencies obtained were not additive. The decreased recombination frequencies observed could reflect: a) the limited degree of base-sequence homology which exists between HSV type 1 and 2 DNAs, and/or b) differences in the genetic organization of the two genomes. If the genomes of HSV type 1 and 2 contain essentially the same genes but in different sequences, it is possible that an intertypic recombination event could lead to the production of lethal hybrids which would not contain a full complement of essential genes. Such recombinants would not be scored and the recombination frequencies would be proportionally lower.

Clearly, additional studies of both *intra-* and *inter*typic complementation and recombination as well as studies of regulation of HSV type 1 and 2 gene expression will be required to elucidate the structural and functional differences in the genomes of the two types of HSV.

VI. Phenotypic Characterization of Herpesvirus *ts* Mutants

In order to define the function which a particular gene performs in the viral replicative cycle, physiological analysis of *ts* mutant phenotypes is conducted at the npT°. A major drawback to the study of mutant phenotypes is the limited number and sensitivity of available techniques for detecting and quantifying a particular phenotypic characteristic. Despite this drawback, a variety of parameters have been used in the characterization of *ts* mutants of herpesviruses, including viral DNA synthesis, induction of enzymes involved in DNA synthesis such as virus-induced thymidine kinase (TK) and DNA polymerase (DP), synthesis of virus-specific polypeptides, thermal stability of the mutant virion, and the synthesis of physical virus particles.

A. Viral DNA

Viral functions can be classified as being dependent upon, or independent of viral DNA synthesis. Therefore, the viral DNA phenotype of mutants at the npT° is often the first parameter to be examined. Since the density of herpesvirus DNAs differs appreciably from that of cellular DNA, viral and cellular DNAs can be sufficiently well separated by ultracentrifugation to determine whether individual mutants synthesize detectable amounts of viral DNA at the npT°. The ability of *ts* mutants of HSV type 1, type 2, and PRV to synthesize viral DNA at pT° and npT° as compared with the WT virus has been determined by analytical ultracentrifugation of DNA in infected cell lysates and by ³H-thymidine labeling of viral and cellular DNA, followed by preparative ultracentrifugation.

SUBAK-SHARPE et al. (1973) determined by preparative ultracentrifugation of ³H-thymidine-labeled cell extracts that 4 HSV type 1 mutants in 3 out of 8 complementation groups were either severely or completely blocked in viral DNA synthesis at the npT° (Table 2). Using a similar technique, HALLIBURTON and TIMBURY (1973) were able to detect viral DNA in lysates of only 3 out of 10 HSV type 2 mutants at the npT° (Table 3). Furthermore, 2 DNA⁻ mutants (*ts*9 and *ts*11) failed to shut off host-cell DNA synthesis and one (*ts*9) did not inhibit histone synthesis at this temperature. In extracts of cells infected with *ts*9 at the npT°, no virus-specific proteins were detected by polyacrylamide gel electrophoresis (PAGE), however an increase in TK activity was observed.

Extensive phenotypic characterization of herpesvirus *ts* mutants has been conducted using the HSV type 1 and type 2 mutants isolated by SCHAFFER et al. (1970, 1973) and ESPARZA et al. (1974), respectively. Table 6 presents a summary of the data available to date on the phenotypes of these mutants. Mutants in 4 of the 15 groups of HSV type 1 (A–D) and in 3 of the 7 groups of HSV type 2 (A–C) were unable to synthesize detectable levels of viral DNA at the npT° as determined by analytical ultracentrifugation (SCHAFFER et al., 1973; ESPARZA et al., 1974) and for HSV type 1 mutants by preparative ultracentrifugation of labeled extracts (ARON et al., in preparation).

ARON et al. (in preparation) noted that mutant DNA phenotypes were identical when either analytical or preparative ultracentrifugation techniques were used, indicating that the two methods can be used interchangeably for routine determinations. These authors also noted that in ³H-thymidine-labeled cultures infected with certain mutants and incubated at the npT°, the yields of infectious *ts* (leak) and *ts*⁺ (revertant) virus reached 10³ PFU/ml. Since no viral DNA was detected in extracts of these cultures, the limit of detectability of the incorporation technique was obviously greater than 10³ infectious genome equivalents/ml.

The only other group of herpesvirus *ts* mutants for which data are available regarding viral DNA synthesis is the group of 10 PRV *ts* mutants reported by PRINGLE et al. (1973). Although none of the mutants incorporated radio-activity into viral DNA as efficiently as the WT virus at either temperature, five mutants, representing 4 out of 9 complementation groups, were severely restricted in the synthesis of viral DNA at the npT° as judged by the in-corporation of both ³H-thymidine and ³²P-orthophosphate. In addition, several mutants defective in viral DNA synthesis were found to inhibit host-cell DNA synthesis to a lesser extent than the WT virus.

Assuming that herpesvirus genes involved in viral DNA synthesis are not highly mutable, the observation that one-third to one-half of the essential functions of herpesviruses so far identified are involved directly or indirectly in viral DNA synthesis is in sharp contrast to results obtained with *ts* mutants of adenovirus, another DNA-containing virus. Wilkie et al. (1973) showed that mutants in only one of 14 complementation groups of adenovirus type 5 were DNA$^-$ and Shiroki et al. (1972) showed that none of 13 complementation groups of adenovirus type 12 were DNA$^-$ at the npT°.

B. Induction of Viral DNA Polymerase (DP)

Attempts have been made to correlate the DNA phenotype of the HSV type 1 and type 2 mutants of Schaffer et al. (1970, 1973) and Esparza et al. (1974) with their ability to induce viral DP at the npT° as determined by assays of DP activity in infected and mock-infected cell extracts (Aron et al., 1973a; Esparza et al., 1974). Enzyme activity was measured 10 to 12 hours after infection at the npT° by a modification of the method of Kit et al. (1967). As shown in Table 6, a close correlation existed between the DNA phenotypes of the mutants and their DP phenotype at the npT°. DNA$^+$ mutants of both HSV type 1 and type 2 induced DP activity at levels approaching that of the WT virus. The DNA$^-$ mutants of HSV type 1 in complementation groups B, C, and D and the DNA$^-$ mutants of HSV type 2 either failed to induce DP at the npT° (DP$^-$) or they exhibited reduced DP activity (DP$^\pm$) as compared with WT virus. Only mutants in HSV type 1 group A synthesized WT levels of DP and yet synthesized no detectable viral DNA.

Tests of DP activity in whole-cell extracts of group A DNA$^-$ HSV type 1 *ts* mutants have demonstrated the DP$^+$ phenotype of these mutants, indicating that viral DNA replication is not essential for the induction of DP activity. Further studies (Aron et al., in preparation) have indicated that infected cell extracts of these mutants were deficient in nuclear but not in cytoplasmic DP activity, suggesting a defect in transport of DP from the cytoplasm to the nucleus. In addition, the deficiency in DP activity of the DNA$^-$ mutants of both HSV types 1 and 2 could not be attributed to a diffusable inhibitor that might have been present in infected cells at the npT° (Aron et al., 1973a; Esparza et al., 1974; Purifoy and Benyesh-Melnick, 1974).

Aron et al. (in preparation) also obtained evidence indicating that the DP induced by the DNA$^-$ mutants of HSV type 1 in groups C and D at the pT° was significantly more thermolabile than the enzyme induced by the WT virus, suggesting a *ts* lesion in the DP molecule *per se*. The observations made up till now on DP activity suggest that at least 4 cistrons of HSV type 1 (A–D) and 3 cistrons of HSV type 2 (A–C) affect the regulation and/or synthesis of both viral DP and viral DNA.

C. Virus-Induced Thymidine Kinase (TK)

Detailed studies of the synthesis of virus-induced TK by HSV type 1 and type 2 mutants at the npT° (ARON et al., 1973b; ESPARZA et al., 1974) have failed to demonstrate: a) a correlation between viral DNA and TK phenotypes, or b) a correlation between the TK phenotypes of mutants and their ability to replicate at the npT°. A summary of these data is presented in Table 6. Studies of virus-induced TK of HSV type 1 *ts* mutants revealed that 6 out of 8 mutants with defects in TK activity were BrdUrd-derived and furthermore, that 5 out of the 6 were TK⁻ at both pT° and npT° (ARON et al., 1973b). These observations suggest that selective mutation of the TK gene by BrdUrd or selection of TK⁻ *ts* mutants may have occurred during mutant induction with this agent. The apparent temperature-sensitivity of the TK induced by one *ts* mutant at the npT° supports the notion that HSV codes for a new TK (KIT and DUBBS, 1969; MUNYON et al., 1972). Furthermore, the observation that mutations in viral TK at the pT° had no effect on viral replication corroborates the findings of KIT and DUBBS (1963) that TK is not essential for virus replication.

In a study of TK induction with 10 PRV *ts* mutants, PRINGLE et al. (1973) demonstrated that all 10 mutants induced an increase in TK activity at the npT°.

D. HSV-Specific Polypeptide Synthesis

As shown in Table 6, regardless of their viral DNA phenotype, all HSV type 1 and type 2 mutants examined were able to induce virus-specific antigens at the npT° detectable by immunofluorescence (R. J. COURTNEY, personal communication; ESPARZA et al., 1974). The ability of these mutants to synthesize virus-specific polypeptides at both pT° and npT° has been examined in detail by SDS-PAGE of nuclear and cytoplasmic fractions of mutant-infected cells (SCHAFFER et al., 1971; BONE et al., 1973; BONE and COURTNEY, 1974; COURTNEY and BENYESH-MELNICK, 1974) (Table 6). Generally speaking the PAGE profiles of DNA⁺ mutants of both HSV type 1 and 2 were similar to that of the WT virus at the npT° (R. J. COURTNEY and K. L. POWELL, personal communication). On the other hand, HSV type 1 DNA⁻ mutants belonging to complementation groups A through D demonstrated significant differences in viral polypeptide synthesis at the npT°. One mutant belonging to group A (*ts*A1) also exhibited a block in the glycosylation of all virus-induced polypeptides at the npT° (SCHAFFER et al., 1971) while infection of cells at this temperature with a second mutant in group A (*ts*A16) resulted in the accumulation of a large molecular weight polypeptide, VP37 (M. W. 37000). Normally, under permissive conditions VP37 can be detected only during the early stages of the infectious cycle (R. J. COURTNEY, personal communication). Infection of cells at the npT° with the group B mutants (*ts*B2 and *ts*B21) resulted in the accumulation of a large molecular weight polypeptide, VP175 (M. W. 175000) which under permissive conditions can

Table 6. Phenotypic characteristics of *ts* mutants of HSV type 1 and type 2

Wild-type Virus	Comple-mentation group	Phenotype at the nonpermissive temperature[a]								
		Viral DNA	DNA polymer-ase	Thymi-dine kinase	Viral antigens (IF)	Viral polypeptides (VP)		Physical particles Naked	Enveloped	Virion thermal stability
						Cytoplasmic	Nuclear			
HSV type 1 (Str. KOS)	A[b]	−	+	+	+	↑ VP37[c]	↑ VP37	−	−	=WT
	B	−	±	±	+	↑ VP175 ↓ VP154, 80	↑ VP175 ↓ VP154, 80	−	−	=WT
	C	−	−	+	+	↑ VP37 ↓ VP154, 80	↑ VP37 ↓ VP154, 80	−	−	< WT(*ts4*) =WT(*ts7*)
	D	−	−	ND	+	=WT	↓ VP154	−	−	=WT
	G H, I, L, M	+	+	+	+	=WT	=WT	−	−	< WT =WT
	E K	+	+	+	+	=WT	=WT	±	−	< WT =WT
	F, J, N, O	+	+	+	+	=WT	=WT	+	±	=WT
HSV type 2 (Str. 186)	A	−	± (*ts1*) − (*ts8*)	+	+	=WT	=WT	±	−	< WT
	B	−	−	+	+	=WT	↑ VP134 ↓ VP154	+	−	= WT
	C	−	±	−	+	=WT	↑ VP134	−	−	=WT
	D	+	+	−	+	=WT	↑ VP134	+	±	< WT
	E, F G	+	+	+	+	=WT	=WT	+	+	< WT =WT

[a] Abbreviations used: ND = not done; WT = wild-type virus; VP values given as molecular weight × 10³.
[b] For mutants in each complementation group, see Table 1.
[c] Mutant *ts1* in this group also exhibited a block in glycosylation of all VP (SCHAFFER et al., 1971).

be transiently detected early after infection. Additional observations with DNA⁻ *ts* mutants of HSV type 1 include: a) reduced synthesis of the major capsid polypeptide VP154 (M. W. 154000), and b) the absence of a "late" polypeptide, VP80 (M. W. 80000) (COURTNEY and BENYESH-MELNICK, 1974). Infection of cells at the npT° with the group C mutants resulted in a block in the synthesis of the major capsid polypeptide, VP154, and in VP80, and an accumulation of VP37. The block in VP154 synthesis occurred before 4 hours post-infection as demonstrated in temperature shift-up experiments. Expression of the *ts* defect with regard to VP154 synthesis at the npT° could be reversed when cultures were shifted down to the pT°, even after 12 hours at the npT°. This reversal occurred in the presence of cytosine arabinoside but not in the presence of actinomycin D. These results thus suggest an early transcriptional or translational defect (BONE and COURTNEY, 1974). The group D mutant (*ts*D9) is characterized by an apparent block in transport of the major capsid polypeptide (VP154) from the cytoplasm to the nucleus at the npT° (R. J. COURTNEY, personal communication).

Regarding viral polypeptide synthesis by DNA⁻ mutants of HSV type 2, cytoplasmic fractions of cells infected with these mutants at the npT° revealed PAGE profiles which were indistinguishable from the profile of the WT virus; however, the nuclear fraction of cells infected with a group B mutant showed reduced amounts of the major capsid polypeptide VP154 and an apparent accumulation of VP134 (M. W. 134000). The latter polypeptide is synthesized in large quantities primarily during the early stages of WT virus replication. Similar accumulations of VP134 were shown to occur in mutants belonging to complementation groups C and D (K. L. POWELL and R. J. COURTNEY, personal communication).

The results of PAGE studies of HSV-induced polypeptides have demonstrated: a) that DNA⁻ mutants of both HSV types 1 and 2 exhibited greater alterations in virus-specific polypeptide synthesis than DNA⁺ mutants, and b) that the defects exhibited by type 2 DNA⁻ mutants were qualitatively and quantitatively less marked than those exhibited by HSV type 1 DNA⁻ mutants. Whether the observed differences in polypeptide synthesis by HSV type 1 and 2 mutants are a consequence of inherent differences in the transcriptional and/or translational programs of the two virus types or to the nature of the individual mutants studied remains to be determined. Further analysis of mutant-infected cell extracts at both the pT° and npT° by high resolution PAGE is currently in progress.

E. Synthesis of Physical Virus Particles

Differences were also observed between the mutants of HSV types 1 and 2 in their ability to assemble viral proteins into physical virus particles as detected by negative staining of cell-free preparations of mutant-infected cultures at the npT° (SCHAFFER et al., 1971, 1974; ESPARZA et al., 1974) (Table 6). While the DNA⁺ mutants of both types exhibited no major differences in polypeptide synthesis, assembly of proteins into virus particles was less efficient among the DNA⁺ HSV type 1 mutants (groups G, H, I, L, M, E, and K) than among the DNA⁺ HSV type 2 mutants (groups D, E, F, and G), suggesting that the defects of the HSV type 1 mutants had a greater effect on maturation. Furthermore, the DNA⁻ mutants of HSV type 1 were more defective in both polypeptide synthesis and particle assembly than the DNA⁻ mutants of HSV type 2, suggesting a more profound consequence of failure to synthesize viral DNA on HSV type 1 than on type 2.

F. Thermal Stability of Mutant Virion Components

Mutants in 3 of 15 HSV type 1 cistrons (C, G and E) and in 4 of 8 HSV type 2 cistrons (A, D, E, and F) were found to synthesize virions at the pT° which were more thermolabile than WT virions when tested at the npT°, suggesting that these mutants possess defects in structural proteins (Schaffer et al., 1973; Esparza et al., 1974) (Table 6). Furthermore, Halliburton and Timbury (1973) found that 5 out of 10 *ts* mutants of HSV type 2 were slightly, but not markedly, more thermolabile than the WT virus. Whether these defects are the primary lesions of the mutants in question remains to be determined. However, since about 50% of the HSV genome codes for structural proteins (Spear and Roizman, 1972), it would not be surprising if a high proportion of HSV *ts* mutants contained primary defects in virion structural components.

G. Electron-Microscopic Observations of HSV Type 1 *ts* Mutants in Thin Sections

By definition temperature-sensitive lesions are expressed at times during the replicative cycle when essential gene products are required. In the absence of an infectious virus, all HSV type 1 *ts* mutants isolated by Schaffer et al. (1973) synthesized viral structural polypeptides and many synthesized viral DNA and physical virus particles at the npT°. The ultrastructural appearance of cells infected with these mutants at the npT° was therefore examined in thin section in order to determine the extent of intracellular morphogenesis (Schaffer et al., 1974b). These studies demonstrated that mutants representing 15 complementation groups exhibited defects in a variety of morphogenic steps at the npT° (Fig. 3). Examples of the particle types and HSV-specific alterations diagramed in Fig. 3 are presented in Fig. 4. While no detectable virus particles were synthesized by class A mutants (in complementation groups B and G, Figs. 3 and 4A), the remaining mutants in classes B and C, 3 of which were DNA⁻, were able to induce the synthesis of variable numbers of physical particles in nuclei of infected cells (Figs. 3 and 4C). While all DNA⁻ and some DNA⁺ mutants synthesized particles which appeared to contain no DNA, only DNA⁺ mutants in class C synthesized dense-cored nucleocapsids, i.e. those which apparently contained viral DNA as illustrated

Mutant Class	Virus	Viral DNA Phenotype 39°	Ring-like Component	Aberrant	Empty and Partial Core	Cylindroid Core	Dense Core	Empty and Partial Core	Dense Core	Membrane Reduplication
				NUCLEUS — Nucleocapsids				**CYTOPLASM AND/OR EXTRACELLULAR SPACE — Enveloped Nucleocapsids**		
	WT	+	±[a]	+	++	−	++	±	++	+
A	tsB2	−	−	−	−	−	−	−	−	−
	tsG3	+	++	−	−	−	−	−	−	±
B	tsC4	−	−	+	+	−	−	−	−	−
	tsD9	−	±	−	+	++	−	±[b]	−	±
	tsA16	−	−	−	++	−	−	±[b]	−	±
	tsH10	+	−	−	+	−	−	−	−	−
	ts111	+	−	−	+	−	−	−	−	−
	tsL14	+	−	−	+	−	−	−	−	±
	tsM19	+	−	−	+	−	−	−	−	±
C	tsK13	+	−	−	+	−	−	±[b]	±[b]	±
	tsE6	+	±	+	++	+	±	±[b]	±[b]	+
	tsF18	+	−	±	++	−	±	±[b]	±[b]	+
	tsO22	+	−	−	++	−	±	+[c]	±[c]	+
	tsN20	+	−	++	++	−	+	±[c]	+[c]	+
	tsJ12	+	−	++	++	−	+	±[c]	+[c]	+

Fig. 3 A–C. Virion forms and virus-specific alterations observed in cells infected with wild-type (WT) virus and *ts* mutants at the npT° (39°) (Table 2; SCHAFFER et al., 1973). (A) Evaluations based on the frequency of virion forms and the degree of alteration observed in *virus-positive* cells only; — absent; ± present in small amounts or numbers; + present in moderate amounts or numbers; + + present in large amounts or numbers. (B) Enveloped particles were observed at the cell surface. (C) Enveloped particles were observed at the nuclear membrane and at the cell surface (after SCHAFFER et al., 1974b)

in Fig. 4H and 4I for WT virus. Several features of thin-sectioning studies are noteworthy:

(1) In *ts*G3-infected cells at the npT°, large accumulations of a ring-like component measuring 250 Å in diameter were noted in the absence of detectable virus particles and other HSV-specific changes (Fig. 4A). These structures have been observed by others who have suggested that they may represent core material (ROIZMAN, 1969; NII, 1971).

(2) The kinds of viral cores observed at the npT° were similar to those observed by others in HSV-infected cells (Fig. 4). In thin sections of two mutants, *ts*D9 and *ts*E6, however, a previously undescribed core form resembling two longitudinally sectioned cylinders was observed (Figs. 4D and 4E). The authors speculated that this core structure may be analogous to the cylindrical component of the toroid around which HSV and MDV DNAs reportedly are spooled (FURLONG et al., 1972; NAZERIAN, 1974).

(3) Although it is generally accepted that primarily viral capsids with dense cores acquire envelopes, moderate numbers of apparently empty particles with envelopes were observed in thin-section preparations of one mutant, *ts*O22, at the npT° (Figs. 4F and 4G). This suggests that envelopment may not depend upon the presence of a full genome equivalent of DNA within the capsid but perhaps upon some other modification of either the viral DNA or a viral structural protein.

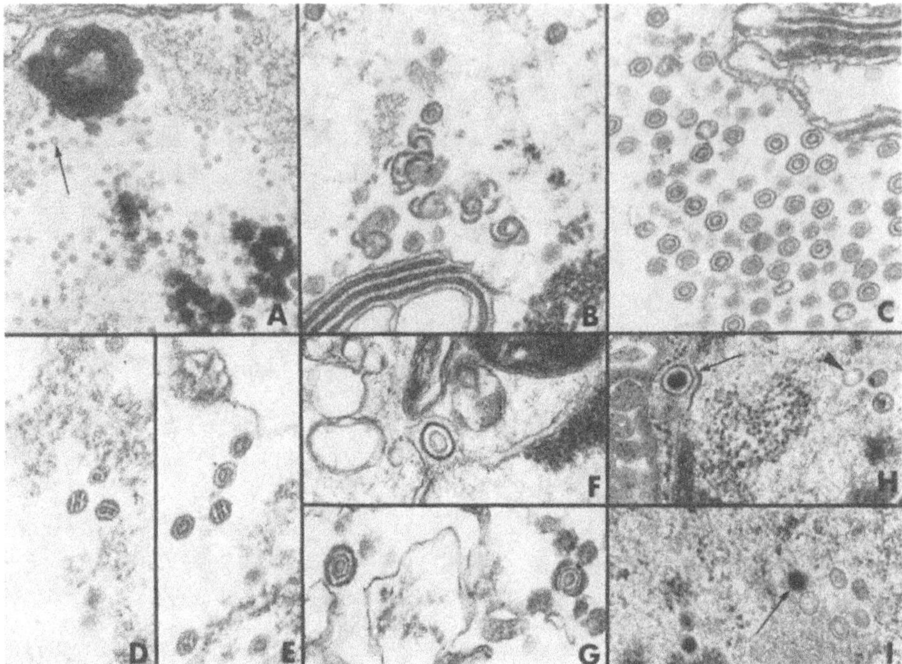

Fig. 4 A–I. Thin sections of HEL cells infected with *ts* mutants and WT virus after 48 hrs at 39° (Table 2; SCHAFFER et al., 1973). (A) Section through the nucleus of a *ts*G3-infected cell showing ring-like structures (arrow, →). (B) Aberrant nucleocapsids in the nucleus of a *ts*N20-infected cell. (C) Section through the nucleus of a *ts*E6-infected cell showing large numbers of partial nucleocapsids. (D) and (E) Thin sections through nuclei of *ts*D9-infected cell showing nucleocapsids containing cylindroid cores. (F) Section through *ts*O22-infected cell showing an enveloped nucleocapsid containing an empty core. (G) Enveloped capsids with partial cores in the extracellular space; *ts*O22-infected cells. (H) WT virus-infected cells; enveloped, dense-cored nucleocapsid in nuclear-cytoplasmic space (arrow, →) and empty nucleocapsid in the nucleus (arrow head, ►). (I) WT virus-infected cells; dense-cored nucleocapsid (arrow, →) and partial nucleocapsids in the nucleus. (× 55000) (after SCHAFFER et al., 1974b)

Studies of the genetic and phenotypic characteristics of HSV and PRV *ts* mutants have so far served to emphasize several features of herpesvirus replication:

From complementation analysis and studies of mutant viral DNA phenotypes at the npT°, it is apparent that a large proportion (1/3 to 1/2) of the essential information encoded by the HSV and PRV genomes represents information for functions which control viral DNA synthesis. It has been reported that both HSV types 1 and 2 code for 24 virus-specific structural polypeptides with a combined molecular weight of about 2580000 daltons (SPEAR and ROIZMAN, 1972). Structural polypeptides thus correspond to approximately 30–40% of the total genetic information of the virus (SPEAR and ROIZMAN, 1972), a fact which accounts for the relatively frequent isolation of mutants with *ts* defects in structural polypeptides. Since a large part of the essential information encoded in herpesvirus DNA controls viral DNA synthesis and a large part of the total genome controls the synthesis of viral structural polypeptides, it would not be surprising if the synthesis of some viral structural polypeptides were involved in the control of viral DNA synthesis and/or *vice versa*. One consequence of such a situation would be the expression of covariant properties. The marked alterations in HSV-specific polypeptide synthesis by DNA$^-$ *ts* mutants of HSV type 1 may reflect such covariation.

Since many of the HSV type 1, type 2, and PRV *ts* mutants are apparently unable to synthesize viral DNA at the npT°, they can be regarded as "early" mutants in the classical sense. However, most DNA$^-$ *ts* mutants of HSV type 2 and PRV are able to synthesize most or all viral polypeptides, including structural polypeptides, in a manner similar to that of the WT viruses. In addition, many are able to synthesize physical virus particles in large numbers at this temperature. The synthesis of structural polypeptides occurs *after* and is dependent upon, viral DNA synthesis in other virus systems, i.e. genes for structural proteins are classically considered to be "late" genes. In the case of DNA$^-$ *ts* mutants of herpesviruses, however, viral DNA synthesis does not appear to be necessary for the synthesis of most viral polypeptides. This would imply that highly efficient transcription and translation occur with DNA$^-$ mutants using the information encoded by the parental genome. Several precedents exist for the synthesis of proteins and viral RNA in the absence of herpesvirus DNA synthesis: when HSV-DNA synthesis was inhibited by hydroxyurea, WAGNER et al. (1972) reported that "early" HSV-RNA was synthesized and NII et al. (1968) found that viral structural proteins were produced and limited assembly of virus particles occurred. The synthesis of PRV structural proteins was not affected when DNA synthesis was inhibited by cytosine arabinoside (BEN-PORAT et al., 1969). Thus, the distinction between "early" and "late" functions with respect to viral DNA and structural protein synthesis in herpesvirus systems is not clear.

VII. Use of *ts* Mutants in Studies of Herpesvirus Replication, Pathogenesis, and Transformation

In addition to information obtained in the routine genetic and biochemical characterization of herpesvirus *ts* mutants, their utilization in studies designed to solve particular problems concerning virus replication and transformation has just begun. A brief review of these studies is presented in the following section.

A. Isolation and Characterization of a Large Molecular Weight Polypeptide of HSV Type 1

The advantages which *ts* mutants offer for studies of the events in virus replicative cycles have been described in detail elsewhere (FENNER, 1970). The availability of two *ts* mutants of HSV type 1 (*ts*B2 and *ts*B21, Tables 2

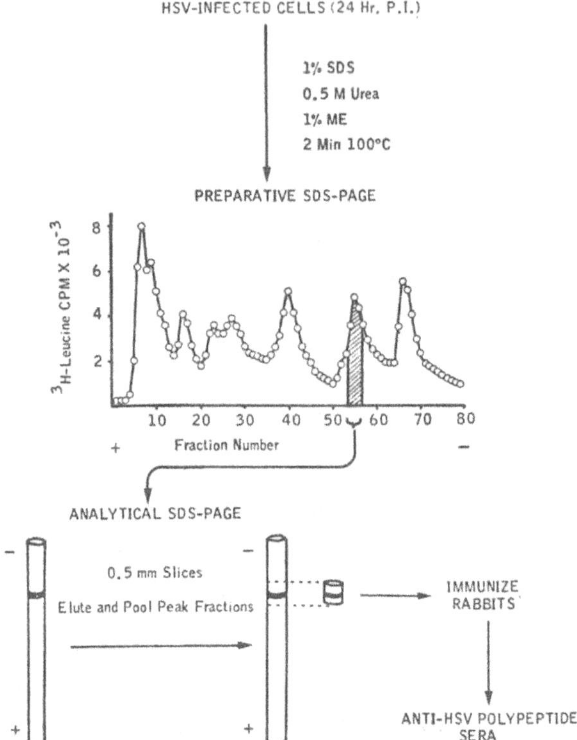

Fig. 5. Method for preparation of HSV polypeptide-specific antiserum (after Courtney and Benyesh-Melnick, 1974)

and 6) has permitted the isolation and preliminary characterization of a virus-induced polypeptide which in WT virus infections could be detected only transiently and in relatively small amounts early in HSV infection (Courtney and Benyesh-Melnick, 1974). This polypeptide, designated VP175 (M. W. 175 000) accumulated in cytoplasmic and nuclear fractions of cells infected with tsB2 at the npT° (Table 6). In contrast, at the pT°, VP175 was detected as a minor component primarily in the nuclear fraction of cells infected with either tsB2 or the WT virus. Taking advantage of this observation, VP175 was isolated from the cytoplasmic fraction of cells infected with tsB2 at the npT° using a combination of SDS-preparative and analytical PAGE as shown in Fig. 5. Antiserum prepared in rabbits to the purified polypeptide was found to be free of neutralizing antibody to HSV type 1 and reacted strongly by immunofluorescence in both nuclei and cytoplasm of tsB2-infected cells at the npT°. At the pT°, however, cells infected with both tsB2 and the WT virus exhibited only nuclear reactivity. Immunofluorescence data thus confirmed the location of VP175 at pT° (nucleus) and npT° (cytoplasm and nucleus) as determined by PAGE analysis. VP175 was shown to be at least partially type-specific since cells infected with the HSV type 1 WT virus exhibited strong nuclear fluorescence while HSV type 2 WT-infected cells showed, at best, a very weak, diffuse cytoplasmic reaction. Preliminary SDS-PAGE analysis indicated that in HSV type 2 WT-infected cells, a polypeptide with a molecular weight comparable to the VP175 observed in HSV type 1 WT virus-infected cells, was also synthesized. Immunofluorescence data, therefore, indicate a possible difference in the compartmentalization of VP175 in HSV type 1 and type 2 infected cells and suggest that this polypeptide may bear primarily type-specific antigenic determinants. Clearly, *ts* mutants in which viral polypeptides accumulate at the npT° should facilitate the isolation, purification, and characterization of these polypeptides and assist in determining the roles which they play in virus replication.

B. Complementation of Adeno-Satellite Virus in Cells Infected with *ts* Mutants of HSV

Adeno-satellite viruses (ASV) are unable to replicate by themselves. Adenoviruses function as complete helpers for ASV replication in that infectious ASV is produced in mixed infections (MAYOR, 1973). HSV and other herpesviruses function as partial helpers in that ASV structural antigens and viral DNA, but no infectious virus, are produced (ATCHISON, 1970; BOUCHER et al., 1971). To elucidate the partial helper function(s) which HSV provides for ASV, production of ASV type 4 structural antigens and infectious DNA was studied in mixed infections with *ts* mutants of HSV types 1 and 2 at the npT° (DRAKE et al., 1974). Like HSV type 1 WT virus, mutants in HSV type 1 DNA⁻ complementation groups A and D, and DNA⁺ group O complemented ASV antigen synthesis at the npT° as detected by immunofluorescence, while members of DNA⁻ groups C and D did not (Table 6). Furthermore, infectious ASV-DNA was isolated from cells mixedly infected with ASV and the HSV type 1 group A DNA⁻ mutant *ts*1. Mutants in HSV type 2 DNA⁻ groups B and C also complemented ASV antigen synthesis at the npT°. The ability of DNA⁻ *ts* mutants of both HSV types 1 and 2 to complement ASV structural antigens and viral DNA demonstrates that the synthesis of these molecules was not dependent upon HSV-DNA synthesis. Consequently, the helper function appears to be one which is expressed before HSV-DNA synthesis. The fact that mutants in two DNA⁻ complementation groups were defective in their ability to complement ASV antigen synthesis at the npT° but not at the pT° indicates that at least two HSV cistrons in some way control ASV antigen synthesis. These mutants should be of value in determining the biochemical nature of the helper functions which HSV provides to ASV.

C. Herpesvirus *ts* Mutants: Virulence and Vaccine Studies

As with other viruses, the acquisition of *ts* characteristics with cut-off temperatures in the range of normal body temperature modifies the *in vivo* behavior of herpesviruses. The decreased virulence of *ts* mutants compared with that of the WT viruses from which they were derived offers an opportunity to examine the role of both virus and host-cell functions in the process of pathogenesis. Studies have so far been made on virulence properties of *ts* mutants of HSV type 1, type 2, and IBR viruses.

In an investigation of the *in vivo* behavior of HSV type 1 *ts* mutants (Tables 1, 2 and 6; SCHAFFER et al., 1970, 1973), the virulence of 16 mutants representing 15 complementation groups was tested by intraperitoneal inoculation of newborn mice (SCHAFFER and WIMBERLY, in preparation). Only 3 DNA$^+$ mutants, *ts*F17, *ts*M19, and *ts*N20, representing 3 complementation groups (Table 2) were shown to be as virulent as the WT virus while all other mutants were significantly less virulent. The range of newborn mouse ipLD$_{50}$ of the WT virus and the most virulent mutants was $< 10^1$–10^4 PFU while that of the less virulent mutants was 10^5–10^6 PFU. While revertant, *ts*$^+$, virus was isolated from the brains and livers of mice which died following inoculation with some mutants, *ts* virus was isolated from others. The sera of mice which survived inoculation with mutant virus contained neutralizing antibody to the WT virus. The level of neutralizing activity detected was dependent upon the mutant DNA phenotype (DNA$^+$ viruses induced higher levels of neutralizing antibody). Four maximally attenuated mutants, *ts*B2, *ts*C7, *ts*D9, and *ts*E6, were able to induce resistance to intracranial challenge with WT virus six weeks after inoculation.

In a second study, ZYGRAICH and HUYGELEN (1973, Table 1) examined the *in vivo* behavior of 3 *ts* mutants of HSV type 2 in rabbits and mice. While either intradermal or intramuscular inoculation failed to induce any neurological symptoms in rabbits, and resulted in a sharp reduction in symptoms and deaths in mice, the WT virus exhibited marked neuropathogenicity. Although no detectable neutralizing antibodies developed following intradermal or intramuscular inoculation of either rabbits or mice with the mutants, most animals exhibited a marked degree of protection to subsequent challenge with the WT virus. The protection was more marked in animals which had received the mutants by the intradermal route.

Infectious bovine rhinotracheitis virus (IBR) causes respiratory disease, genital symptoms, meningoencephalitis, conjunctivitis, and is able to induce abortion. In an effort to test the potential of *ts* mutants of IBR for vaccine purposes, ZYGRAICH et al. (1974a, 1974b) examined the *in vitro* and *in vivo* behavior of the IBR *ts* mutant, RLB 106 (Table 1). The mutant proved to be genetically stable *in vitro*. Intranasal inoculation of susceptible calves with RLB 106 produced no fever or other symptoms, yet nearly 100% of the animals seroconverted. The vaccinated calves exhibited: a) protection to subsequent challenge with the WT virus, b) reduced excretion of virus, and c) a typical booster response. During prolonged direct contact between vaccinated and unvaccinated animals, transmission occurred but no symptoms were detected in unvaccinated controls. Vaccination with *ts*-RLB 106 induced high levels of interferon which persisted for several days post-vaccination in the noses of inoculated calves. In addition, the virus was shown to be non-abortigenic. The work of ZYGRAICH et al. (1974a, 1974b) represents the first in-depth study of a herpesvirus *ts* mutant for use as a potential vaccine. The degree of attenuation, immunogenicity, genetic stability, and the fact that the *ts* property of the mutant is an easily recognizable virological marker satisfy the criteria for a potential vaccine.

Although these studies have demonstrated the usefulness of HSV type 1, type 2, and IBR *ts* mutants for studies of viral pathogenesis, the use of potentially oncogenic herpesviruses for vaccine purposes should be undertaken with caution. The *in vitro* oncogenic potential of herpesviruses to be tested for possible use as vaccines should be thoroughly investigated. Vaccines composed of nucleic acid-free viral subunits may prove to be more suitable for vaccine purposes. In this regard, it may be possible to obtain large quantities of individual viral structural components from cultures infected with certain *ts* mutants in which these components accumulate at the npT° (Table 6) (COURTNEY and BENYESH-MELNICK, 1974; SCHAFFER et al., 1974).

D. HSV Type 2 *ts* Mutants as Transforming Agents

Initial attempts to transform cells *in vitro* with HSV type 1, type 2, and cytomegalovirus have utilized UV- or photo-inactivated virus (DUFF and RAPP, 1971, 1973; ALBRECHT and RAPP, 1973; RAPP et al., 1973). Both methods were designed to inactivate viral lytic functions leaving transforming functions intact. It is likely, however, that UV- and photo-inactivation of lytic genes also resulted in the inactivation of transforming functions which may account for the low efficiency of transformation observed in these studies. In attempts to induce transformation using *ts* mutants at the npT°, synthesis of infectious virus is inhibited by virtue of the *ts* lesion, while the remaining viral genes are left intact. Using this technique, MACNAB (1974) successfully induced transformation of rat embryo cells with TIMBURY's (1971) *ts* mutants of HSV type 2 at the npT°. Positive transformation, as indicated by HSV-specific surface immunofluorescence and extended passage life of transformed cells, was obtained with two DNA⁻ and three DNA⁺ *ts* mutants at the npT°. Similar results have been obtained by KIMURA et al. (in preparation) using a DNA⁻ *ts* mutant of HSV type 2, *ts*B5 (Table 3). In neither study, however, was there any indication that transformation with *ts* mutants was more efficient than other methods.

A primary objective of studies of viral transformation is to determine which viral genes are involved in the transformation process. Therefore, to demonstrate the temperature-sensitivity of the transformed cell phenotype would strongly suggest that the *ts* viral gene was involved in the maintenance of the transformed state. Although this principle has been demonstrated successfully with both DNA- and RNA-containing oncogenic viruses (TOOZE, 1973), whether the HSV *ts* mutant-transformed cells described above are temperature-sensitive with respect to transformation markers is not known.

E. Detection of HSV Gene Functions in Cells Transformed by HSV Using *ts* Mutants

Hamster cells transformed by UV-irradiated HSV type 2 contain viral antigens detectable by immunofluorescence techniques (DUFF and RAPP, 1971), and approximately 10% of HSV-specific RNA was found to be transcribed in these cells (COLLARD et al., 1973). In an attempt to determine whether gene products which are essential for HSV replication were produced and were biologically active in HSV type 2-transformed cells, KIMURA et al. (1974a, 1974b) examined the replication of 8 *ts* mutants of HSV type 2 (Tables 1 and 2; ESPARZA et al., 1974) in these cells at the npT°. The growth of 2 DNA$^+$ *ts* mutants with defects in two different cistrons (*ts*D6 and *ts*G4) was enhanced from 13 to 36 times in 2 HSV type 2-transformed cell lines (333-8-9 and MS4-1; DUFF and RAPP, 1971) at the npT° as compared with their growth in normal hamster cells. The replication of the remaining 6 *ts* mutants (4 DNA$^-$ and 2 DNA$^+$) was not enhanced more than 4-fold in these cell lines. Furthermore, no enhancement of yields of any of the 8 mutants was observed in SV40-induced tumor cells.

Enhancement of HSV type 2 *ts* mutant growth in type 2-transformed cells suggests that functional HSV type 2 gene products assisted the replication of the mutants at the npT° by complementation mechanisms. On the other hand, the absence of enhancement of mutant replication in SV40-transformed cells at this temperature implies: a) that homologous genetic information is necessary for enhancement, and b) that an oncogenically transformed cell phenotype *per se* is insufficient to produce enhancement. It is of interest that the two mutants (*ts*D6 and *ts*G4) which exhibited enhanced growth are located in the same region of the HSV type 2 linkage map (Fig. 1 C). This clustering could be coincidental or, alternatively, this region of the HSV type 2 genetic map may code for gene products involved in some manner in the HSV type 2 transformation process. The fact that the replication of the same two mutants was enhanced in two independently derived HSV type 2-transformed cell lines indicates perhaps some specificity of the HSV type 2 information functioning in HSV type 2-transformed cells. In any event, the demonstration of enhanced growth of type 2 *ts* mutants in type 2-transformed cells suggests the efficacy of undertaking similar studies in, e.g. cervical cancer cells and in other herpesvirus systems for which *ts* mutants of the transforming viral agent are available. The identification of the functional defects expressed by the mutants and supplied by the resident partial or complete genome must then be identified and this is not a simple matter. Having demonstrated the presence of functional viral information in virus-transformed cells, it may also be possible to rescue this information by recombination at the pT° with a superinfecting *ts* mutant.

VIII. Herpesvirus *ts* Mutants: Future Investigations

The studies with herpesvirus *ts* mutants just described have demonstrated the suitability of these viruses for conventional genetic analysis including the isolation of mutants, the analysis of their patterns of complementation and recombination, and the construction of genetic maps. However, the analysis of an animal virus genome composed of nearly 100 genes using classical genetic techniques constitutes an overwhelming task. Recent studies of the structural and functional organization of the SV40 and adenovirus genomes have utilized unique viral DNA fragments, generated by bacterial restriction endonucleases, in combination with *ts* mutants and deletion mutants for genetic analysis. This approach permits the physical mapping of the entire viral genome and provides a means for relating specific viral functions to a physical location on the viral chromosome. The method is based on the striking enzymatic specificity of purified restriction endonucleases for viral DNAs and on the separation of the cleavage products by gel electrophoresis.

The use of DNA fragments for genetic analysis represents a variation of deletion mutant mapping. While deletion mutants *lack* a portion of the viral genome, viral DNA fragments *contain* only a portion of the genome. A major advantage of the use of fragments in the genetic analysis of herpesviruses would be that investigative efforts could be focused on the analysis of a small segment of the large herpesvirus genome. The fragments of viral DNA produced by restriction enzymes are ordered either by isolating partial digest products and digesting them to completion, thus identifying overlapping clusters of fragments, or by sequential digestion using enzymes with different specificities (DANNA et al., 1973). Knowing the order of fragments on the physical map, marker rescue of *ts* mutants by individual fragments permits the *ts* mutational sites in individual cistrons to be localized to specific segments of the genome (LAI and NATHANS, 1974). By mapping a large number of *ts* mutants, including representatives of all complementation groups, it may be possible to define the limits of each cistron.

Prerequisites for mapping studies with DNA fragments include: (a) a technique for reproducibly cleaving viral DNA into a number of unique fragments, (b) a method for insuring the quantitative uptake of viral DNA and DNA fragments in a biologically active form, and (c) a series of *ts* mutants with defects in a variety of cistrons. The necessary techniques for mapping the complete herpesvirus genome are now available, and indeed, these studies are currently in progress. SKARE et al. (submitted for publication) and N. M. WILKIE (personal communication) fragmented the DNAs of several strains of HSV type 1 and 2 using restriction endonucleases. SHELDRICK et al. (1973) and GRAHAM et al. (1973) demonstrated biologically active (infectious) HSV-DNA. Several series of HSV type 1 and 2 *ts* mutants are currently available for physical mapping studies as described above. N. M. WILKIE (personal communication), using *ts* mutants and the *syn* plaque morphology marker, reported preliminary evidence for marker rescue from fragmented DNA and correlation of the fragments to the known position of the *ts* markers on the linkage map of BROWN et al. (1973). These studies thus demonstrate the feasibility of constructing a complete fragmentation map of the HSV genome showing the location of all viral cistrons for which *ts* mutants are available. Although the order of markers and the distances between them should be consistent on linkage maps and on fragmentation maps, linkage map distances are based on recombination frequencies, which represent the *relative* distance only between two markers, and not the actual physical distance. Therefore, despite the fact that herpesvirus linkage maps have been constructed, we do not know whether these maps represent the entire viral genome or only a portion of it. Fragmentation mapping, on the other hand, can define the limits of existing linkage maps and delineate those segments of the genome for which no *ts* markers have yet been identified.

Since only a few of the total number of cistrons of any herpesvirus have so far been identified, the isolation of additional mutants is clearly necessary in order to obtain mutants in the remaining essential genes. In addition, efforts should be made to determine the total number of cistrons so far identified by exchange of mutants between laboratories.

Although studies of the genetic interactions between two *ts* mutants and between *ts* mutants and DNA fragments are possible without knowing the nature of the *ts* lesion exhibited by each mutant, functional significance can be ascribed to each viral cistron only by exhaustive phenotypic characterization of mutants. A primary limitation imposed on the phenotypic characterization of *ts* mutants is the small number of herpesvirus gene functions which have been clearly identified and which can be assayed quantitatively. Furthermore, investigations of *ts* mutants may not reveal the functions of all virus genes, since it is not certain that conditional-lethal mutations can occur in any gene. Thus, EPSTEIN et al. (1963) noted that the occurrence of *ts* mutations in certain genes of bacteriophage was a very rare event. In addition, there are herpesvirus genes which code for proteins whose functions are not essential for virus replication, e.g. viral TK.

Transformation of cells using discrete, well-characterized fragments of herpesvirus DNA offers several advantages for studies of viral transformation and for the identification of transforming genes: a) Inactivation by heat or UV-irradiation of viral transforming and replicative genes (see Section *VII D* above) would not present a problem if DNA fragments were used as transforming agents. The lethal effects of intact genomes would also be avoided. b) The use of DNA fragments limits, by exclusion, the number of genes being tested. Rather than an intact genome, groups of genes can be examined separately for their transforming ability; yet the entire complement of viral genes can be tested. Transformation by a particular fragment would demonstrate that the genes encoded by that fragment were sufficient in themselves to induce transformation. F. L. GRAHAM (personal communication) and N. M. WILKIE (personal communication), working with adenovirus and HSV, respectively, obtained evidence that viral DNA fragments can transform rat cells *in vitro*. c) Genes which induce nononcogenic transformation can theoretically be separated from genes which induce oncogenic transformation.

As indicated earlier, a method for detecting specific viral functions in herpesvirus-transformed cells was reported by KIMURA et al. (1974a). The use of *ts* mutants as probes for demonstrating herpesvirus functions can also be applied to cells potentially transformed *in vivo*, e.g. human cervical cancer cells. Whether the same viral genes which are essential for replication are essential in virus transformation, however, is unknown. The *ts* mutants currently available are defective in essential replicative functions by definition.

Since infectious virus is inducible in cells latently infected with herpesvirus, the entire viral genome must be present. Clearly, however, all genes present are not functional. The use of *ts* mutants as probes for viral gene functions can also be applied to cells latently infected with herpesviruses.

The *ts* mutants described in this review should be regarded as tools for understanding the structural and functional properties of herpesvirus genomes in virus-infected and virus-transformed cells. The identification of specific functional genes in these cells is only a first step in investigations of herpesvirus replication, latency, and transformation. Problems of regulation of herpesvirus gene function in latently infected and transformed cells—where viral gene expression is under cellular control--will require further investigation.

Acknowledgements. I should like to express my thanks to Dr. I. HALLIBURTON, University of Leeds, England; Dr. C. HUYGELEN, Recherche et Industrie Therapeutiques, Rixensart, Belgium; Dr. S. KATO, Osaka University, Japan; Drs. M. TIMBURY and C. PRINGLE, MRC Institute of Virology, Glasgow, Scotland; and Dr. M. TERNI, Universita degli Studi, Ferrara, Italy, for kindly making their data, both published and unpublished, available to me; I should also like to thank my colleagues in the Department of Virology, in particular Drs. G. ARON, N. BISWAL, D. BONE, R. COURTNEY, J. ESPARZA, S. KIMURA, and D. PURIFOY, whose work forms the basis for many of the studies cited in this paper. I am especially indebted to Drs. J. BUTEL, R. COURTNEY, and M. BENYESH-MELNICK, for helpful comments and discussions, and to Ms. C. DUNN for her secretarial assistance. The published and unpublished data from our laboratory cited here were accumulated with the aid of funding from The Virus Cancer Program and other grants from the National Cancer Institute, National Institutes of Health.

References

ALBRECHT, T., RAPP, F.: Malignant transformation of hamster embryo fibroblasts following exposure to ultraviolet-irradiated human cytomegalovirus. Virology **55**, 53–61 (1973)

ARON, G. M., SCHAFFER, P. A., BENYESH-MELNICK, M.: DNA polymerase activity in DNA-negative temperature-sensitive mutants of herpes simplex virus. Abstracts Ann. Meet. Amer. Soc. Microbiol. V **42**, 201 (1973a)

ARON, G. M., SCHAFFER, P. A., COURTNEY, R. J., BENYESH-MELNICK, M., KIT, S.: Thymidine kinase activity of herpes simplex virus temperature-sensitive mutants. Intervirology **1**, 96–109 (1973b)

ATCHISON, R. W.: The role of herpes virus in adenovirus-associated virus replication *in vitro*. Virology **42**, 155–162 (1970)

AURELIAN, L.: Virions and antigens of herpes virus type-2 in cervical carcinoma. Cancer Res. **33**, 1539–1547 (1973)

Bachenheimer, S. L., Kieff, E. D., Lee, L. F., Roizman, B.: Comparative studies of DNAs of Marek's disease and herpes simplex viruses. In: Oncogenesis and herpesviruses, Biggs, P. M., de-The, G., Payne, L. N. (eds.). Lyon: Int. Agency Res. Cancer 1972

Becker, Y., Dym, H., Sarov, I.: Herpes simplex virus DNA. Virology **36**, 184–192 (1968)

Ben-Porat, T., Shimono, H., Kaplan, A.: Synthesis of proteins in cells infected with herpesvirus. II. Flow of structural viral proteins from cytoplasm to nucleus. Virology **37**, 56–61 (1969)

Biswal, N., Murray, B. K., Benyesh-Melnick, M.: Ribonucleotides in newly synthesized DNA of herpes simplex virus. Virology **61**, 87–99 (1974)

Bone, D. R., Benyesh-Melnick, M., Courtney, R. J.: A temperature-sensitive mutant of herpes simplex virus defective in the synthesis of the major capsid polypeptide. Abstracts Ann. Meet. Amer. Soc. Microbiol. V 320, 247 (1973)

Bone, D. R., Courtney, R. J.: A temperature-sensitive mutant of herpes simplex virus type 1 defective in the synthesis of the major capsid polypeptide. J. gen. Virol. **24**, 17–27 (1974)

Boucher, W., Melnick, J. L., Mayor, H. D.: Nonencapsidated infectious DNA of adenosatellite virus cells coinfected with herpesvirus. Science **173**, 1243–1245 (1971)

Brown, S. M., Ritchie, D. A., Subak-Sharpe, J. H.: Genetic studies with herpes simplex virus type 1. The isolation of temperature-sensitive mutants, their arrangement into complementation groups and recombination analysis leading to a linkage map. J. gen. Virol. **18**, 329–346 (1973)

Burge, B. W., Pfefferkorn, E. R.: Isolation and characterization of conditional-lethal mutants of Sindbis virus. Virology **30**, 204–213 (1966a)

Burge, B. W., Pfefferkorn, E. R.: Complementation between temperature-sensitive mutants of Sindbis virus. Virology **30**, 214–223 (1966b)

Cameron, K. R.: Ultraviolet irradiation of herpes simplex virus. Action spectrum for the survival of infectivity in relation to the small-plaque effect. J. gen. Virol. **18**, 51–54 (1973)

Cerdá-Olmedo, E., Honawalt, P. C., Guerola, N.: Mutagenesis of the replication point by nitrosoguanidine: Map and pattern of replication of the *Escherichia coli* chromosome. J. molec. Biol. **33**, 705–719 (1968)

Chou, J. Y., Martin, R. C.: Complementation analysis of simian virus 40 mutants. J. Virol. **13**, 1101–1109 (1974)

Collard, W., Thornton, W. H., Green, M.: Cells transformed by human herpesvirus type 2 transcribe virus-specific RNA sequences shared by herpesvirus types 1 and 2. Nature (Lond.) New Biol. **243**, 264 (1973)

Cooper, P.: The mutation of poliovirus by 5-fluorouracil. Virology **22**, 186–192 (1964)

Cooper, P. D.: The genetic analysis of poliovirus. In: The biochemistry of viruses, Levy, H. B. (ed.). New York: Dekker 1969

Courtney, R. J., Benyesh-Melnick, M.: Isolation and characterization of a large molecular weight polypeptide of herpes simplex virus type 1. Virology **62**, 539–551 (1974)

Danna, K. J., Sack, G. H., Jr., Nathans, D.: A cleavage map of the SV40 genome. J. molec. Biol. **78**, 363–376 (1973)

Drake, J. W.: The molecular basis of mutation. San Francisco-London-Cambridge-Amsterdam: Holden-Day 1970

Drake, S., Schaffer, P. A., Esparza, J., Mayor, H. D.: Complementation of adeno-associated satellite viral antigens and infectious DNA by temperature-sensitive mutants of herpes simplex virus. Virology **60**, 230–236 (1974)

Duff, R., Rapp, F.: Oncogenic transformation of hamster cells after exposure to herpes simplex virus type 2. Nature (Lond.) New Biol. **233**, 48 (1971)

Duff, R., Rapp, F.: Oncogenic transformation of hamster embryo cells after exposure to inactivated herpes simplex virus type 1. J. Virol. **12**, 209–217 (1973)

Epstein, R. H., Bolle, A., Steinberg, C. M., Kellenberger, E., Edgar, R. S., Susman, M., Denhardt, G. H., Lielausis, A.: Physiological studies of conditional lethal mutants of bacteriophage T4D. Cold Spr. Harb. Symp. quant. Biol. **28**, 375–394 (1963)

Erikson, R. L., Szybalski, W.: The Cs_2SO_4 equilibrium density gradient and its application for the study of T-even phage DNA: Glucosylation and replication. Virology **22**, 111–124 (1964)

Esparza, J.: Studies with temperature-sensitive mutants of herpes simplex virus type 2. Ph. D. Dissertation, Baylor College of Medicine, 1974

Esparza, J., Purifoy, D. J. M., Schaffer, P. A., Benyesh-Melnick, M.: Isolation, complementation and preliminary phenotypic characterization of temperature-sensitive mutants of herpes simplex virus type 2. Virology 57, 554–565 (1974)

Farley, C. A., Banfield, W. G., Kasnic, G., Foster, W. S.: Oyster herpes-type virus. Science 178, 759–760 (1972)

Fenner, F.: Conditional lethal mutants of animal viruses. Curr. Topics Microbiol. Immunol. 48, 1–28 (1969)

Fenner, F.: The genetics of animal viruses. Ann. Rev. Microbiol. 24, 297–334 (1970)

Fields, B. N., Joklik, W. K.: Isolation and preliminary genetic and biochemical characterization of temperature-sensitive mutants of reovirus. Virology 37, 335–342 (1969)

Figueroa, M. E., Rawls, W. E.: Biological markers for differentiation of herpesvirus strains of oral and genital origin. J. gen. Virol. 4, 259–267 (1969)

Fitch, W. M.: Does the fixation of neutral mutations form a significant part of observed evolution in proteins? In: Evolution of genetic systems, Smith, H. H. (ed.). New York: Gordon & Breach 1972

Freese, E.: Molecular mechanism of mutations, In: Molecular genetics, Taylor, J. M. (ed.). New York: Academic Press 1963

Frenkel, N., Roizman, B.: Separation of the herpesvirus deoxyribonucleic acid duplex into unique fragments and intact strand on sedimentation in alkaline gradients. J. Virol. 10, 565–572 (1972)

Furlong, D., Swift, H., Roizman, B.: Arrangement of herpesvirus deoxyribonucleic acid in the core. J. Virol. 10, 1071–1074 (1972)

Ghendon, Y. Z.: Conditional lethal mutants of animal viruses. Progr. med. Virol. 14, 68–122 (1972)

Gordin, M., Olshevsky, U., Rosenkranz, H. S., Becker, Y.: Studies on herpes simplex virus DNA: Denaturation properties. Virology 55, 280–284 (1973)

Graham, B. J., Ludwig, H., Bronson, D. L., Benyesh-Melnick, M., Biswal, N.: Physicochemical properties of the DNA of herpes viruses. Biochim. biophys. Acta (Amst.) 259, 13–23 (1972)

Graham, F. L., Veldhuisen, G., Wilkie, N. M.: Infectious herpesvirus DNA. Nature (Lond.) New Biol. 245, 265–266 (1973)

Halliburton, I. W., Timbury, M. C.: Characterization of temperature-sensitive mutants of herpes simplex virus type 2. Growth and DNA synthesis. Virology 54, 60–68 (1973)

Hay, J., Perera, P. A. J., Morrison, J. M., Gentry, G. A., Subak-Sharpe, J. H.: In: Strategy of the viral genome, Wolstenholme, G. G. W., O'Connor, M. (eds.). Edinburgh: Churchill Livingstone. Ciba Foundation Symposium 1971

Hayes, W.: The genetics of bacteria and their viruses. Studies in basic genetics and molecular biology, 2nd ed. New York: John Wiley 1968

Hirsch, I., Vonka, V.: Ribonucleotides linked to DNA of herpes simplex virus type 1. J. Virol. 13, 1162–1168 (1974)

Honess, R. W., Roizman, B.: Proteins specified by herpes simplex virus. XI. Identification and relative molar rates of synthesis of structural and nonstructural herpes virus polypeptides in the infected cell. J. Virol. 12, 1347–1365 (1973)

Huang, E. S., Pagano, J. S.: Human cytomegalovirus. II. Lack of relatedness to DNA of herpes simplex I and II, Epstein-Barr virus, and nonhuman strains of cytomegalovirus. J. Virol. 13, 642–645 (1974)

Hunt, R. D., Melendez, L. V.: Herpes virus infections of non-human primates: A review. Lab. Animal Care 19, 221–234 (1969)

Huy, V. D., Stäber, H., Waschke, K., Rosenthal, H. A.: Isolation and complementation of ts mutants of pseudorabies virus. In: International virology, Melnick, J. L. (ed.), vol. 2. Basel: Karger 1972

Kazama, F. Y., Schornstein, K. L.: Herpes-type virus particles associated with a fungus. Science 177, 696–697 (1972)

Kieff, E., Hoyer, B., Bachenheimer, S., Roizman, B.: Genetic relatedness of type 1 and type 2 herpes simplex viruses. J. Virol. 9, 738–745 (1972)

Kimura, S., Esparza, J., Benyesh-Melnick, M., Schaffer, P. A.: Enhanced replication of temperature-sensitive mutants of herpes simplex virus type 2 (HSV-2) at the nonpermissive temperature in cells transformed by HSV-2. Intervirology, 3, 162–169 (1974 b)

KIMURA, S., SCHAFFER, P. A., BENYESH-MELNICK, M.: Replication of *ts* mutants of herpes simplex virus type 2 (HSV-2) at the nonpermissive temperature in cells transformed by HSV-2. Abstracts Amer. Assoc. Cancer Res., p. 58 (1974a)

KIRKWOOD, J., GEERING, G., OLD, L. J.: Demonstration of group and type-specific antigens of herpesviruses. In: Oncogenesis and herpesviruses, BIGGS, P. M., DE-THE, G., PAYNE, L. N. (eds.). Lyon: Int. Agency Res. Cancer 1972

KIT, S., DUBBS, D. R.: Acquisition of thymidine kinase activity by herpes simplex infected mouse fibroblast cells. Biochem. biophys. Res. Commun. **11**, 55–59 (1963)

KIT, S., DUBBS, D. R.: Enzyme induction by viruses. In: Monographs in virology, MELNICK, J. L. (ed.), vol. 2. Basel-New York: Karger 1969

KIT, S., PIEKARSKI, L. J., DUBBS, D. R.: DNA polymerase induced by simian virus 40. J. gen. Virol. **1**, 163–173 (1967)

KLEIN, G.: Herpesviruses and oncogenesis. Proc. nat. Acad. Sci. (Wash.) **69**, 1056–1064 (1972)

KOMENT, R. W., RAPP, F.: Conditional lethal mutants of herpes simplex virus type 2. Abstracts Ann. Meet. Amer. Soc. Microbiol. V **200**, 234 (1974)

LAI, C.-J., NATHANS, D.: Mapping temperature-sensitive mutants of simian virus 40: Rescue of mutants by fragments of viral DNA. Virology **60**, 466–475 (1974)

LANNI, Y. T., LANNI, F., TEVETHIA, M. J.: Bacteriophage T5 chromosome fractionation: Genetic specificity of a DNA fragment. Science **152**, 208–210 (1966)

LOW, M., HAY, J., KEIR, H. M.: DNA of herpes simplex virus is not a substrate for methylation *in vivo*. J. molec. Biol. **46**, 205–207 (1969)

LOW, M., MECHIE, M. I., HAY, J.: Methylation of pseudorabies-virus deoxyribonucleic acid. Biochem. J. **124**, 63P (1971)

LUDWIG, H. O., BISWAL, N., BENYESH-MELNICK, M.: Studies on the relatedness of herpesviruses through DNA-DNA hybridization. Virology **49**, 95–101 (1972)

MACNAB, J. C. M.: Transformation of rat embryo cells by temperature sensitive mutants of herpes simplex virus. J. gen. Virol. **24**, 143–153 (1974)

MANSERVIGI, R.: Method for isolation and selection of temperature-sensitive mutants of herpes simplex virus. Appl. Microbiol. **27**, 1034–1040 (1974)

MAYOR, H. D.: Satellite viruses. Meth. Cancer Res. **8**, 203–236 (1973)

MUNYON, W., BUCHSBAUM, R., PAOLETTI, E., MANN, J., KRAISELBURD, E., DAVIS, D.: Electrophoresis of thymidine kinase activity synthesized by cells transformed by herpes simplex virus. Virology **49**, 683–689 (1972)

NAHMIAS, A. J., CHANG, G. C. H., FRITZ, M. E.: Herpesviruses as infectious and oncogenic agents in man and other vertebrates. In: Membranes and viruses in immunopathology, DAY, S. B., GOOD, R. A. (eds.). New York: Academic Press 1972

NAHMIAS, A. J., DOWDLE, W. R.: Antigenic and biologic differences in herpesvirus hominis. Progr. med. Virol. **10**, 110–159 (1968)

NAHMIAS, A. J., JOSEY, W. E., NAIB, Z. M., LUCE, C. F., GUEST, B. A.: Antibodies to *herpesvirus hominis* types 1 and 2 in humans. II. Women with cervical cancer. Amer. J. Epidemiol. **91**, 547–552 (1970)

NAZERIAN, K.: DNA configuration in the core of Marek's disease virus. J. Virol. **13**, 1148–1150 (1974)

NII, S.: Electron microscopic observations on FL cells infected with herpes simplex virus. I. Viral forms. Biken's J. **14**, 177–190 (1971)

NII, S., ROSENKRANZ, H. S., MORGAN, C., ROSE, H. M.: Electron microscopy of herpes simplex virus. III. Effect of hydroxyurea. J. Virol. **2**, 1163–1171 (1968)

ONODA, T., KONOBE, T., TAKAKU, K., NAITO, M., ONO, K., KATO, S.: Temperature-sensitive growth of a herpes type virus isolated from a chicken with Marek's disease. Biken's J. **14**, 357–360 (1971)

PADGETT, B. L., TOMKINS, J. K. N.: Conditional lethal mutants of rabbit-pox virus. III. Temperature-sensitive (ts) mutants; physiological properties, complementation and recombination. Virology **36**, 161–167 (1968)

PLUMMER, G., GOODHEART, C. R., HANSON, D., BOWLING, C. P.: A comparative study of the DNA density and behavior in tissue cultures of fourteen different herpesviruses. Virology **39**, 134–137 (1969)

PRINGLE, C. R.: Genetic characteristics of conditional lethal mutants of vesicular stomatitis virus induced by 5-fluorouracil, 5-azacytidine and ethyl methane sulfonate. J. Virol. **5**, 559–567 (1970)

PRINGLE, C. R., HOWARD, D. K., HAY, J.: Temperature-sensitive mutants of pseudorabies virus with differential effects on viral and host DNA synthesis. Virology **55**, 495–505 (1973)

PURIFOY, D. J. M., BENYESH-MELNICK, M.: DNA polymerase induction by DNA-temperature-sensitive (ts) mutants of herpes simplex virus type 2 (HSV-2). Abstracts Ann. Meet. Amer. Soc. Microbiol. V **201**, 234 (1974)

RAPP, F., LI, J.-L. H., JERKOFSKY, M.: Transformation of mammalian cells by DNA-containing viruses following photodynamic inactivation. Virology **55**, 339–346 (1973)

RAWLS, W. E., TOMPKINS, W. A. F., FIGUEROA, M. E., MELNICK, J. L.: Herpesvirus type 2: Association with carcinoma of the cervix. Science **161**, 1255–1256 (1968)

ROBB, J. A., MARTIN, R. G.: Genetic analysis of simian virus 40. I. Description of microtitration and replica-plating techniques for virus. Virology **41**, 751–760 (1970)

ROIZMAN, B.: The herpesviruses—A biochemical definition of the group. Curr. Topics Microbiol. Immunol. **49**, 1–79 (1969)

ROIZMAN, B., FRENKEL, N.: The transcription and state of herpes simplex virus DNA in productive infection and in human cervical cancer tissue. Cancer Res. **33**, 1402–1416 (1973)

ROIZMAN, B., SPEAR, P. G., KIEFF, E. D.: Herpes simplex viruses I and II: A biochemical definition. Perspect. Virol. **8**, 129–169 (1972)

ROIZMAN, B., et al.: Provisional labels for herpesviruses. Report of the Herpesvirus Study Group, International Committee for the Nomenclature of Viruses. J. gen. Virol. **20**, 417–419 (1973)

ROSS, L. J. N., CAMERON, K. R., WILDY, P.: Ultraviolet irradiation of herpes simplex virus: Reactivation processes and delay in virus multiplication. J. gen. Virol. **16**, 299–311 (1972b)

ROSS, L. J. N., FRAZIER, J. A., BIGGS, P. M.: An antigen common to some avian and mammalian herpesviruses. In: Oncogenesis and herpesviruses, BIGGS, P. M., DE-THE, G., PAYNE, L. N. (eds.). Lyon: Int. Agency Res. Cancer 1972a

RUSSELL, W. C., CRAWFORD, L. V.: Properties of the nucleic acids from some herpes group viruses. Virology **22**, 288–292 (1964)

SAVAGE, T., ROIZMAN, B., HEINE, J. W.: Immunological specificity of the glycoproteins of herpes simplex virus subtypes 1 and 2. J. gen. Virol. **17**, 31–48 (1972)

SCHAFFER, P. A., ARON, G. M., BISWAL, N., BENYESH-MELNICK, M.: Temperature-sensitive mutants of herpes simplex virus type 1: Isolation, complementation and partial characterization. Virology **52**, 57–71 (1973)

SCHAFFER, P. A., BRUNSCHWIG, J. P., McCOMBS, R. M., BENYESH-MELNICK, M.: Electron microscopic studies of temperature-sensitive mutants of herpes simplex virus type 1. Virology **62**, 444–457 (1974b)

SCHAFFER, P. A., COURTNEY, R. J., McCOMBS, R. M., BENYESH-MELNICK, M.: A temperature-sensitive mutant of herpes simplex virus defective in glycoprotein synthesis. Virology **46**, 356–368 (1971)

SCHAFFER, P. A., LAM, K., WIMBERLY, I., BENYESH-MELNICK, M.: Temperature-sensitive mutants of *herpesvirus saimiri*. Manuscript in preparation

SCHAFFER, P. A., TEVETHIA, M. J., BENYESH-MELNICK, M.: Recombination between temperature-sensitive mutants of herpes simplex virus type 1. Virology **58**, 219–228 (1974a)

SCHAFFER, P., VONKA, V., LEWIS, R., BENYESH-MELNICK, M.: Temperature-sensitive mutants of herpes simplex virus. Virology **42**, 1144–1146 (1970)

SHELDRICK, P., LAITHIER, M., LANDO, D., RYHINER, M. L.: Infectious DNA from herpes simplex virus: Infectivity of double-stranded and single-stranded molecules. Proc. nat. Acad. Sci. (Wash.) **70**, 3621–3625 (1973)

SHIROKI, K., ARISAWA, J., SHIMOJO, H.: Isolation and a preliminary characterization of temperature-sensitive mutants of adenovirus 12. Virology **49**, 1–11 (1972)

SIM, C., WATSON, D. H.: The role of type-specific and cross-reacting structural antigens in the neutralization of herpes simplex virus types 1 and 2. J. gen. Virol. **19**, 217–233 (1973)

SKARE, J., SUMMERS, W. T., SUMMERS, W. C.: Structure and function of herpesvirus genomes. I. Comparison of five HSV-1 and two HSV-2 strains by cleavage of their DNA with *Eco* R1 restriction endonuclease. Submitted for publication

Soehner, R. L., Gentry, G. A., Randall, C. C.: Some physicochemical characteristics of equine abortion virus nucleic acid. Virology 26, 394–405 (1965)

Spear, P. G., Roizman, B.: Proteins specified by herpes simplex viruses. V. Purification and structural proteins of the herpesvirion. J. Virol. 9, 143–159 (1972)

Stephenson, J. R., Reynolds, R. K., Aaronson, S. A.: Isolation of temperature-sensitive mutants of murine leukemia virus. Virology 48, 749–756 (1972)

Subak-Sharpe, J. H.: Base doublet frequency patterns in the nucleic acid and evolution of viruses. Brit. med. Bull. 23, 161–168 (1967)

Subak-Sharpe, J. H.: The genetics of herpesvirus. Cancer Res. 33, 1385–1392 (1973)

Subak-Sharpe, J. H., Brown, S. M., Ritchie, D. A., Timbury, M. C., Halliburton, I. W.: Herpes virus genetics. Advanc. Biosci. 11, 205–218 (1973)

Tevethia, M. J., Ripper, L. W., Tevethia, S. S.: A simple qualitative spot complementation test for temperature-sensitive mutants of SV40. Intervirology, 3, 245–255 (1974)

Thouless, M. E.: Serological properties of thymidine kinase produced in cells infected with type 1 or type 2 herpesvirus. J. gen. Virol. 17, 307–315 (1972)

Timbury, M. C.: Temperature-sensitive mutants of herpes simplex virus type 2. J. gen. Virol. 13, 373–376 (1971)

Timbury, M. C., Subak-Sharpe, J. H.: Genetic interactions between temperature-sensitive mutants of types 1 and 2 herpes simplex viruses. J. gen. Virol. 18, 347–357 (1973)

Timbury, M. C., Theriault, A., Elton, R. A.: A stable syncytial mutant of herpes simplex type 2 virus. J. gen. Virol. 23, 219–224 (1974)

Tooze, J. (ed.): The molecular biology of tumour viruses. Cold Spring Harbor, New York: Cold Spring Harbor Laboratory 1973

Visconti, N., Delbrück, M.: The mechanism of genetic recombination in phage. Genetics 38, 5–33 (1953)

Wagner, E. K., Swanstrom, R. I., Stafford, M. G.: Transcription of the herpes simplex virus genome in human cells. J. Virol. 10, 675–682 (1972)

Wagner, E. K., Tewari, K. K., Kolodner, R., Warner, R. C.: The molecular size of herpes simplex virus type 1 genome. Virology 57, 436–447 (1974)

Watson, D. H., Wildy, P., Harvey, B. A. M., Shedden, W. I. H.: Serological relationships among the viruses of the herpes group. J. gen. Virol. 1, 139–141 (1967)

Wildy, P.: Recombination with herpes simplex virus. J. gen. Microbiol. 13, 34–46 (1955)

Wildy, P.: Classification and nomenclature of viruses. Monogr. Virol. 5, 33–34 (1971)

Wildy, P.: Herpes: History and classification. In: The herpesviruses, Kaplan, A. S. (ed.). New York and London: Academic Press 1973

Wilkie, N. M.: The synthesis and substructure of herpesvirus DNA: The distribution of alkali-labile single strand interruptions in HSV-1 DNA. J. gen. Virol. 21, 453–467 (1973)

Wilkie, N. M., Ustacelebi, S., Williams, J. F.: Characterization of temperature sensitive mutants of adenovirus type 5: Nucleic acid synthesis. Virology 51, 499–503 (1973)

Witter, R. L., Nazerian, K., Purchase, H. G., Burgoyne, G. H.: Isolation from turkeys of a cell-associated herpesvirus antigenically related to Marek's disease virus. Amer. J. vet. Res. 31, 525 (1970)

Zur Hausen, H., Schulte-Holthausen, H.: Presence of EB virus nucleic acid homology in a "virus-free" line of Burkitt tumour cells. Nature (Lond.) 227, 245–248 (1970)

Zygraich, N., Huygelen, C.: In vivo behavior of a temperature-sensitive (ts) mutant of herpesvirus hominis type 2. Arch. ges. Virusforsch. 43, 103–111 (1973)

Zygraich, N., Lobmann, M., Vascoboinic, E., Berge, E., Huygelen, C.: *In vivo* and *in vitro* properties of a temperature-sensitive mutant of infectious bovine rhinotracheitis virus. Res. Vet. Sci., in press (1974b)

Zygraich, N., Vascoboinic, E., Huygelen, C.: Replication of a temperature-sensitive mutant of infectious bovine rhinotracheitis virus in the tissues of inoculated calves. Zbl. Vet.-Med. 21, 138–144 (1974a)

Inhibition of the Multiplication of Enveloped Viruses by Glucose Derivatives[1]

C. SCHOLTISSEK[2]

With 2 Figures

Table of Contents

I. Introduction

Some 15 years ago KILBOURNE (1959) described the effect of 2-deoxy-D-glucose on the multiplication of an influenza B virus in chick embryos. The sugar antimetabolite was applied immediately after infection. In a single cycle experiment the early yield (8 h after infection) of virus was markedly depressed while later (24 h after infection) there was no significant difference compared with the controls. Thus there was a retardation of virus multiplication caused by the sugar antimetabolite. Since this effect could be counteracted by pyruvate, KILBOURNE suggested that the inhibition might be due to an interference with cellular glycolysis. The effect of 2-deoxy-D-glucose was completely reversible *in vitro*, indicating that the host cell was not killed by the deoxy sugar. Although this observation revealed a definite specificity of the sugar antimetabolite on virus multiplication, the effect was not studied further until 1972. In this year two groups independently found that with glucose derivatives like 2-deoxy-D-glucose or glucosamine, the

1 This work was supported by the Sonderforschungsbereich 47 Virologie.
2 Institut für Virologie, Justus Liebig-Universität, Gießen, Germany.

synthesis of influenza glycoproteins was specifically inhibited, while the yield of carbohydrate-free viral proteins was not significantly affected (KALUZA et al., 1972; GANDHI et al., 1972). Since that time the two sugar derivatives have been investigated in many different virus–host systems in order to study the role of the carbohydrate component of glycoproteins in the synthesis and function of these complex molecules. The results described might be of even more general interest, since the same principles found with viral glycoproteins might apply also to cellular glycoproteins. The advantage of the virus–cell systems is that in most cases after infection the synthesis of cellular proteins is abolished so that only one or a few newly induced virus-specific glyco-proteins are produced which can be studied more easily.

Quite a body of results and information has accumulated so that it now seems worthwhile to summarize the data available and to draw some con-clusions.

II. Effect of Glucosamine and 2-Deoxy-D-Glucose on the Host Cells

Glucosamine has been shown to have a generally toxic effect on the metab-olism of animal cells by interfering with DNA, RNA, and protein synthesis (BEKESI et al., 1969). 2-Deoxy-D-glucose is known to interfere with the energy supply of cells by inhibiting glycolysis and respiration, and in this way drastically decreases the ATP pool (WOODWARD and HUDSON, 1954; IBSEN et al., 1962; LETNANSKY, 1964). Thus doses of these sugar derivatives which exert the effects mentioned above are of little interest with regard to their inhibition of virus multiplication, because death of the host cell automatically means inhibition of virus multiplication. One very important point in all these studies is that the cell metabolism should be observed very carefully in order to recognize any non-specific effects on virus production. Therefore this chapter will be concerned with the action of these sugar derivatives on cell metabolism under conditions which inhibit virus multiplication while exerting no or only insignificant effects on host-cell macromolecule synthesis.

A. Glucosamine

Glucosamine is known to deplete the host cells of uridine triphosphate (UTP) according to the following scheme (BEKESI and WINZLER, 1969; SCHOLTISSEK, 1971):

$$\text{Glucosamine} \xrightarrow{\text{ATP}} \text{glucosamine-6-phosphate} \rightarrow$$

$$\rightarrow \text{N-acetyl-glucosamine-6-phosphate} \rightleftharpoons$$

$$\xrightarrow{\text{N-acetyl-glucosamine-1,6-diphosphate}} \text{N-acetyl-glucosamine-1-phosphate} \rightleftharpoons$$

$$\xrightleftharpoons[\text{pyrophosphate}]{\text{UTP}} \text{UDP-N-acetyl-glucosamine.}$$

UDP-N-acetyl-glucosamine accumulates in these cells. According to this scheme ATP is also necessary for the reaction, but since in general the ATP pool is about 10 times larger than the UTP pool, glucosamine has a much smaller effect on the ATP pool. Furthermore, glucosamine also affects the *de novo* synthesis of CTP, because the CTP-synthetase seems to have a lower affinity for UTP compared to the enzyme forming UDP-N-acetyl-glucosamine (uridine diphosphoglucose pyrophosphorylase) (SCHOLTISSEK, 1972). Finally, glucosamine stimulates the uptake of pyrimidine nucleosides into the host cells by lowering the UTP pool (SCHOLTISSEK, 1972). Similarly, galactosamine is able to deplete rat liver cells of UTP by forming UDP-galactosamine (KEPPLER et al., 1970). In chick embryo cells, however, galactosamine does not affect the UTP pool (SCHOLTISSEK, 1971), which is probably due to a rather low activity of galactokinase in these cells.

Since amino sugars seem to inhibit virus multiplication via the effect on the uridine-phosphate derivatives of the host cells, it is important to know something about the corresponding enzyme activities of the host cells under investigation. Pool sizes and/or enzyme activities also change with the age of primary cell cultures. During incubation of primary chick embryo cultures the cell population and cell density changes until a completely contact-inhibited cell layer is reached. Thus the effect of glucosamine on the UTP pool is much less pronounced in chick embryo cells 24 h after seeding compared with cells tested 48 h after seeding (SCHOLTISSEK, 1971; SCHOLTISSEK et al., 1974a).

Another very important factor is the medium used for the propagation of cells in culture. Glucosamine seems to be transported and phosphorylated by the same system as is glucose (PLAGEMANN and ERBE, 1973); therefore the glucose present in all culture media competes with the uptake and metabolism of the amino sugar. For this reason, culture media have been developed in which glucose is replaced by either pyruvate or fructose. The latter two compounds can be used as energy sources, but do not compete with glucosamine. Thus, using media containing pyruvate or fructose (SCHOLTISSEK et al., 1975) only one tenth of the concentration of glucosamine which was required with glucose-containing media is needed in order to obtain the same effect on the UTP pool and on the incorporation of labelled uridine into RNA. This point is illustrated for pyruvate-containing media in Table 1. Increasing doses of glucosamine (added 90 min prior to the pulse with labelled uridine) cause an increase of radioactivity in the fraction of activated sugars (UDP-X) and a decrease of radioactivity within the UTP fraction. Total uptake of label into the acid-soluble pool and into RNA also increases. Since the uptake of uridine into the cells is regulated by negative feedback from the UTP pool (SCHOLTISSEK, 1972) and the increase in radioactivity in RNA has to be interpreted as an increase of the specific radioactivity of the final precursor UTP, the data in Table 1 are an indirect measure of a decrease of the UTP pool caused by the amino sugar.

Table 1. Effect of glucosamine on uridine uptake and incorporation using glucose- and pyruvate-containing Earle's medium

Energy source in Earle's medium	Glucos-amine mM	Dpm $\times 10^4$ in: Total acid soluble pool	UTP	UDP-X	RNA	UTP/ UDP-X
10 mM glucose	0	663	390	259	97	1.50
	0.08	742	426	301	96	1.42
	1	918	441	461	139	0.96
	10	1 000	334	653	158	0.51
10 mM pyruvate	0	758	452	288	104	1.57
	0.08	736	359	360	133	1.00
	1	846	257	571	182	0.45
	10	695	116	568	252	0.20

Glucosamine as listed in the table was added to Earle's medium containing either 10 mM pyruvate or 10 mM glucose 48 h after seeding the primary chick embryo cells and 90 min prior to the pulse with 10 μC ^{3}H-uridine per culture. The length of the pulse was 40 min. For further details see Scholtissek (1971). The small decrease in total uptake of labelled uridine at the highest concentration of glucosamine in pyruvate medium is due to the fact that under these conditions there is already a small effect on the ATP pool. Radioactivity in UMP or UDP was negligible.

B. 2-Deoxy-D-Glucose

2-Deoxy-D-glucose was believed to be metabolized in animal cells only to the 6-phosphate level (e.g. Kipnis and Cori, 1959; Renner et al., 1972), though the nucleoside diphosphate-activated 2-deoxy-D-glucose derivatives have been found in yeast (Fischer and Weidemann, 1964; Biely and Bauer, 1968). Therefore, 2-deoxy-D-glucose has been used in numerous studies on glucose transport in animal cells. More recently, however, 2-deoxy-D-glucose has been shown to be incorporated into viral (Kaluza et al., 1973) as well as into cellular glycoproteins (Steiner et al., 1973) and glycolipids (Steiner and Steiner, 1973). The deoxy-sugar could be recovered undegraded from these macromolecules by mild acid hydrolysis. Recently the activated 2-deoxy-D-glucose derivatives, UDP-2-deoxy-D-glucose and GDP-2-deoxy-D-glucose, have been discovered in a variety of primary and permanent cells in tissue cultures after addition of the deoxy-sugar to the culture medium (Schmidt et al., 1974). Thus 2-deoxy-D-glucose is normally metabolized and activated in mammalian cells and it is only a matter of the technique of the extraction procedure to detect the highly unstable intermediates.

It should be further noted that since GDP-2-deoxy-D-glucose is also produced in these cells, the deoxy-sugar probably functions not only as an antimetabolite to glucose, but also to mannose. (The structures of 2-deoxy-D-glucose and 2-deoxy-D-mannose are identical.) In fact, the incorporation of labelled 2-deoxy-D-glucose into viral glycoproteins is competed more efficiently by mannose than by glucose (Kaluza et al., 1973).

Table 2. Effect of 2-deoxy-D-glucose on uridine uptake and incorporation using glucose- and pyruvate-containing Earle's medium

Energy source in Earle's medium	2-Deoxy-D-glucose mM	Dpm $\times 10^4$ in: Total acid soluble pool	UTP	UDP-X	UDP	RNA
10 mM glucose	0	663	390	259	8	97
	0.16	743	443	270	22	89
	1	648	357	228	56	93
	10	666	247	282	130	107
10 mM pyruvate	0	758	452	288	10	104
	0.16	581	257	202	117	147
	1	204	40	105	55	33
	10	173	32	73	66	21

The same experiment as in Table 1 is illustrated, except that 2-deoxy-D-glucose was used instead of glucosamine. The drastic decrease in total uptake of labelled uridine at 1 mM 2-deoxy-D-glucose and above in pyruvate-containing medium is due to the fact that under these conditions there is already a significant effect on the ATP pool. Such doses, however, have never been investigated in studies to show inhibition of virus multiplication. Radioactivity in UMP was negligible.

2-Deoxy-D-glucose is taken up by animal cells and is phosphorylated by the same system as for glucose (KIPNIS and CORI, 1959; RENNER et al., 1972; KALUZA et al., 1973). Therefore the same considerations concerning the choice of the medium as mentioned above for glucosamine hold for 2-deoxy-D-glucose.

The degree to which 2-deoxy-D-glucose uses UTP by forming UDP-2-deoxy-D-glucose can be demonstrated easily by labelling cells incubated in the presence of the deoxy-sugar with (^{3}H)-uridine. In the untreated control cells, only traces of labelled UDP are present in the acid extract, while in the extract of treated cells, large quantities of labelled UDP are found (Table 2). The radioactivity in UDP is derived from UDP-2-deoxy-D-glucose which is extremely labile at low pH (BIELY and BAUER, 1967) and decomposes into UDP and 2-deoxy-D-glucose. As shown in Table 2 the effect of the deoxy-sugar on the UTP pool is about 50 times greater in pyruvate medium compared with medium containing glucose.

III. Effect of Glucosamine and 2-Deoxy-D-Glucose on Viral Glycoprotein Synthesis

Glucosamine and 2-deoxy-D-glucose interfere with the multiplication of a variety of viruses containing glycoproteins in their envelope (KALUZA et al., 1972; GANDHI et al., 1972; FLEMING, 1973; COURTNEY et al., 1973; LUDWIG et al., 1974; HUNTER et al., 1974). Surprisingly, with certain other glycoprotein-containing enveloped viruses they have relatively little effect on multiplication as long as a glucose-containing culture medium is used (KALUZA et al., 1972; GANDHI et al., 1972). Table 3 summarizes the results which were obtained in

Table 3. Inhibition of the multiplication of enveloped animal viruses by 2-deoxy-D-glucose and glucosamine using glucose-containing media. A survey

Virus	Host	Inhibitor	Inhibition	Reference
Influenza B	Chick embryo	2-deoxy-D-glucose	+	Kilbourne, 1959
Influenza A (Fowl plague, A_0/Bel)	Primary chick embryo cells	2-deoxy-D-glucose	+	Kaluza et al., 1972 Gandhi et al., 1972
Influenza A (Fowl plague)	Primary chick embryo cells	glucosamine	+	Kaluza et al., 1972
Influenza A (Fowl plague, A_0/Be)	HeLa cells	glucosamine	+	Kaluza et al., 1972 Gandhi et al., 1972
Influenza A (A_0/Bel)	HeLa cells	2-deoxy-D-glucose	+	Gandhi et al., 1972
Toga viruses (Semliki Forest and Sindbis)	Primary chick embryo cells	glucosamine	+	Kaluza et al., 1972 Fleming, 1973 Kaluza et al., 1973
Toga viruses (Semliki Forest and Sindbis)	BHK or HeLa cells	glucosamine	+	Kaluza et al., 1972
Toga viruses (Semliki Forest and Sindbis)	Primary chick embryo cells	2-deoxy-D-glucose	+	Kaluza et al., 1972
Parainfluenza (NDV)	Primary chick embryo cells	glucosamine	−	Kaluza et al., 1972 Scholtissek et al., 1975
Rhabdo viruses	Primary chick embryo cells	2-deoxy-D-glucose	− (+)	Kaluza et al., 1972
Rhabdo viruses	Primary chick embryo cells	glucosamine	−	Kaluza et al., 1972
Herpes simplex	BSC 1 cells	2-deoxy-D-glucose	+	Courtney et al., 1973
Pseudorabies	Rabbit kidney cells	2-deoxy-D-glucose	+	Ludwig et al., 1974
Avian sarcoma PR RSV-B and B77-C virus	Secondary chick embryo cells	glucosamine	+	Hunter et al., 1974 Lewandowski et al., 1974

the various virus–host systems. All influenza and toga viruses tested, herpes simplex and pseudorabies viruses, and avian sarcoma viruses were inhibited in their multiplication, while there was little effect on the multiplication of infectious Newcastle disease (NDV) or vesicular stomatitis virus (VSV) under identical conditions. These results were obtained in a variety of different host cells. The degree of inhibition, however, seems to depend to a certain extent on the age of the culture, at least, when primary chick embryo cells were investigated. In cells used 24 h after seeding the inhibitory effect was much less pronounced than with cells 48 h after seeding (SCHOLTISSEK and ROTT, unpublished).

FLEMING (1973) investigated different strains of Semliki Forest virus and reported that the action of glucosamine to a certain degree is strain dependent. In addition, this author found that in sugar-free medium mannosamine also inhibits the multiplication of Semliki Forest virus.

In glucose-containing medium the synthesis of viral RNA or of the carbohydrate-free polypeptides of influenza or of Semliki Forest virus is not impeded by the sugar derivatives (KALUZA et al., 1972; GANDHI et al., 1972; KLENK et al., 1972; KALUZA et al., 1973). Only the production of viral glycoproteins is disturbed. This observation has been utilized in order to determine the stability and precursor relationships of labelled RNA of NDV, influenza, and Semliki Forest virus by depleting the infected cells of UTP during the chase period (SCHOLTISSEK et al., 1972).

When the influenza proteins synthesized in the presence of inhibiting doses of the sugar-derivatives were analyzed by polyacrylamide gel electrophoresis it was found that in place of the hemagglutinin (HA) a new virus-specific polypeptide (HA_0) appeared, which had a similar amino-acid composition to the native glycoprotein. It seems to consist of an under- or unglycosylated polypeptide, because radioactively labelled HA_0, following a chase period of several hours in glucosamine-free medium, migrates in polyacrylamide gels very much like HA. Thus, the HA_0 after completion of the polypeptide chain still seems to function—at least to a certain extent—as a substrate for the sugar transferases (KLENK et al., 1972). A careful analysis of the dose response, the pool size of the activated sugars, and the molecular weight of the hemagglutinin of fowl plague virus (influenza A) has revealed that the carbohydrate content of this viral glycoprotein becomes progressively smaller with increasing doses of glucosamine or 2-deoxy-D-glucose in the culture medium (Fig. 1) (SCHWARZ and KLENK, 1974).

The unglycosylated hemagglutinin HA_0 behaves similarly to the native glycoprotein during its synthesis, migration from the rough to the smooth membranes, and its cleavage to HA_1 and HA_2. The only difference lies in its cleavage point: the cleavage products HA_{01} and HA_{02} are not only smaller but also much more heterogeneous compared with the native split products (KLENK et al., 1972; KLENK et al., 1974; SCHWARZ and KLENK, 1974). The fact that after infection under inhibiting conditions no hemagglutinating activity is induced in these cells implies that the under- or unglycosylated products are not able to agglutinate erythrocytes.

Since no significant neuraminidase activity is induced in glucosamine- or 2-deoxy-D-glucose-treated cultures, it might be assumed that an unimpaired carbohydrate moiety of neuraminidase is necessary to obtain an enzymatically active molecule. This conclusion is somewhat contradictory to the observation that after pronase or trypsin digestion the carbohydrate-containing part of the enzyme can be removed without loss of enzyme activity (ROTT et al., 1972; LAZDINS et al., 1972; R. T. C. HUANG, personal communication). It is

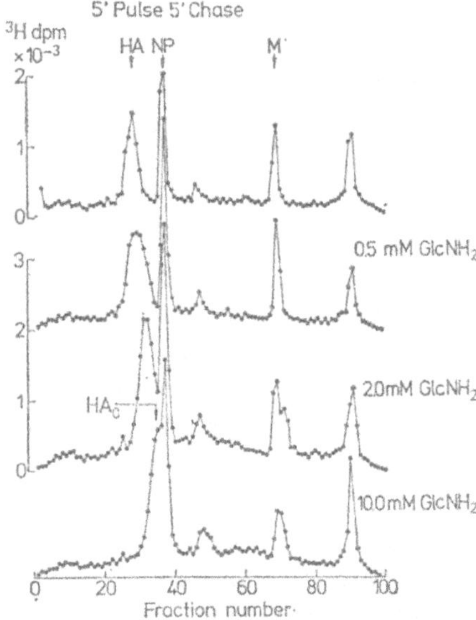

Fig. 1. Influence of glucosamine on the production of HA₀ of fowl plague virus. No, 0.5 mM, 2 mM, or 10 mM glucosamine were added to glucose-containing Earle's medium immediately after infection of chick embryo cells with fowl plague virus. A 5 min pulse with ³H-amino acids (leucine, valine, tyrosine, 10 µC/ml, each) was performed 4 h after infection; the cell extract was analyzed in polyacrylamide gel electrophoresis according to Laemmli (1970) (migration from left to right). (With a 5 min pulse of the glycoproteins only the uncleaved hemagglutinin (HA) is labelled, but not the cleavage products HA₁ and HA₂.) N = nucleocapsid protein, M = matrix protein. (By the courtesy of Dr. H.-D. Klenk)

not known yet, however, how much of the protein together with the carbohydrate moiety is removed by the enzymic digestion.

After incubation of infected chick embryo cells with doses of glucosamine which strongly inhibit the production of influenza glycoproteins and abolish infectivity, no physical virus particles could be isolated from the supernatant medium. When smaller doses were investigated, however, particles containing incorrectly glycosylated hemagglutinin were released from infected cells (Klenk, personal communication).

In chick embryo cells infected with Semliki Forest virus using glucose-containing medium, the nucleocapsid proteins is produced in normal yields in the presence of 2-deoxy-D-glucose. However, in the region of the electropherogram where the non-structural glycoprotein NSP 63 or the two envelope glycoproteins E₁ and E₂ should appear, a great number of new unusual viral polypeptides with different molecular weights were found (Fig. 2) (Kaluza et al., 1973). Furthermore, a polypeptide with a molecular weight of 94 000 accumulates under inhibiting conditions. Similar observations were made when another inhibitor of glycoprotein synthesis, N-fluoroacetyl-glucosamine, was investigated (Morser and Burke, 1974). These results have been interpreted in the sense that glycosylation of a primary gene product in Semliki

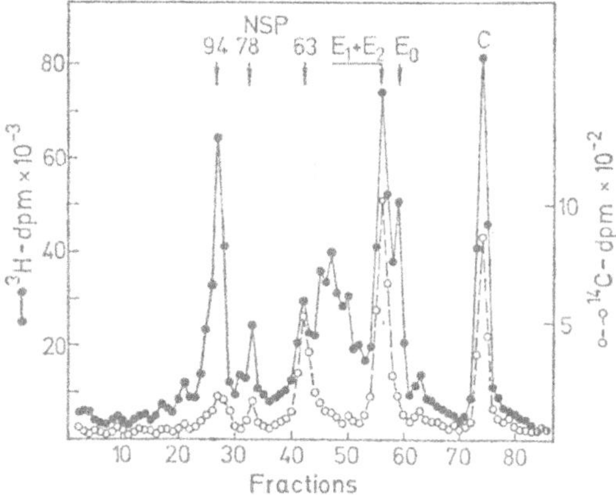

Fig. 2. Effect of 2-deoxy-D-glucose on the synthesis of Semliki Forest virus proteins. Chick embryo cells were incubated after infection with Semliki Forest virus in glucose-containing Earle's medium in the presence of 2 µg/ml of actinomycin D and 0.3 mM 2-deoxy-D-glucose. A 60 min pulse with ³H-labelled leucine, lysine, and tyrosine (a total of 16.7 µCi/ml) was started 4.5 h after infection. For comparison, ¹⁴C-labelled, infected cells incubated without deoxy-sugar were coelectrophoresed. The numbers represent the molecular weights of the non-structural precursor proteins (NSP, daltons ×1 000). E_1 plus E_2 are the two envelope glycoproteins which are not separated in this system. E_0 marks the position where the unglycosylated envelope glycoproteins are expected to appear. C is the capsid protein which is a carbohydrate-free viral protein. (By the courtesy of Dr. G. KALUZA)

Forest virus-infected cells (with an estimated molecular weight of 165 000) is essential for the processing to the final virion polypeptides. Incorporation of labelled 2-deoxy-D-glucose into Semliki Forest virus glycoproteins stops almost immediately after inhibition of protein synthesis by cyclohexamide (KALUZA et al., 1973). From this observation it is concluded that glycosylation occurs mainly in *statu nascendi* onto the growing polypeptide chains or immediately after its completion. Thus if the glycosylation is inhibited or somehow disturbed, under- or unglycosylated intermediates accumulate or/and cleavage occurs at incorrect sites leading to a heterogeneous population of virus-specific polypeptides containing little or no carbohydrate. Under these conditions the yield of virus particles found in the medium is negligible (G. KALUZA, unpublished). This means that the wrongly processed viral polypeptides cannot be used for the formation of (non-infectious) virus particles.

In mannose-containing medium the multiplication of Semliki Forest virus cannot be inhibited by 2-deoxy-D-glucose. The effect of the deoxy-sugar can be counteracted best by mannose, to some extent by glucose (mainly because of competition with the uptake and phosphorylation of the antimetabolite) and not at all by other sugars tested. Thus 2-deoxy-D-glucose seems to act rather as an antimetabolite to mannose than to glucose (KALUZA et al., 1973).

When glucosamine is added to chick fibroblasts transformed by avian sarcoma viruses, virus multiplication stops immediately and no virus particles are formed (HUNTER et al., 1974). Immunofluorescent studies indicate that the glucosamine block affects the production of viral envelope antigens which are glycoproteins. It is now known that corresponding unglycosylated envelope polypeptides are produced under these conditions, since these carbohydrate-free products have been detected using immuno-precipitation techniques (Lewandowski et al., 1974). Synthesis of a group specific protein P 27, which does not contain carbohydrates, is also suppressed by glucosamine, but to a lesser degree than that of the viral envelope antigens. It is not excluded that the production of P 27 somehow depends on the production of a correctly glycosylated common precursor.

Herpesvirus multiplication also can be inhibited by 2-deoxy-D-glucose (Courtney et al., 1973; Ludwig et al., 1974). Incorrect viral glycoproteins are produced in the presence of this sugar antimetabolite. In contrast to influenza and Semliki Forest virus, however, the number of physical particles is not reduced to the same degree as the infectivity is. Instead the particles contain incorrect glycoproteins, and they are, therefore, not infectious (Courtney et al., 1973).

The inhibitory effect of 2-deoxy-D-glucose and/or glucosamine on the multiplication of the enveloped viruses mentioned above is completely reversible at any time during the infectious cycle by removal of the inhibitor and further incubation in fresh medium free of the inhibitor. This fact and the observation that other enveloped viruses (NDV and VSV) undergo almost normal multiplication under identical conditions imply that glucosamine and 2-deoxy-D-glucose do not exert their inhibitory action by an unspecific toxic effect on the host cells.

Now the question arises why the multiplication of some enveloped viruses is inhibited by glucosamine and 2-deoxy-D-glucose under the conditions described, while that of others, e.g. NDV and VSV, is not (Kaluza et al., 1972; Gandhi et al., 1972). One possible explanation is that the uptake and metabolism of the inhibitors are enhanced after infection of the host cells by some viruses, while other viruses have not such an effect. This hypothesis has been studied using 2-deoxy-D-glucose (Scholtissek et al., 1974a). It has been found that the incorporation of the labelled antimetabolite into acid-insoluble material is enhanced severalfold at the time of viral synthesis of two toga viruses (Semliki Forest virus and Sindbis virus) and of an influenza virus, while during synthesis of NDV and VSV there is no such stimulation. Thus there is a clear correlation between the stimulation of the incorporation of the antimetabolite and its inhibiting action. The hypothesis that the two sugar derivatives have to be highly metabolized in order to be effective is strengthened by the observation that after double infection of chick embryo cells by fowl plague virus and VSV, the multiplication of the latter can indeed by inhibited by glucosamine and 2-deoxy-D-glucose (Table 4). At the same time the metabolism of the sugar derivatives is correspondingly enhanced (Table 5) (Rott and Scholtissek, unpublished). There are as yet no data available as to how far the synthesis of the VSV glycoprotein and carbohydrate-free viral protein is impaired by the sugar derivatives under the conditions of double infection.

Table 4. Inhibition of the multiplication of VSV by glucosamine and 2-deoxy-D-glucose
in double infected cells

	Plaque-forming units of VSV
VSV alone	5.3×10^7
VSV + glucosamine	3.5×10^7
VSV + 2-deoxy-D-glucose	2.7×10^7
Fowl plague + VSV	1.0×10^7
Fowl plague + VSV + glucosamine	$8 \quad \times 10^5$
Fowl plague + VSV + 2-deoxy-D-glucose	$6 \quad \times 10^5$

Tissue cultures and infection as in Table 5. Immediately after infection with VSV, 10 mM of the respective sugar derivative was added to Earle's medium containing 10 mM glucose. 6 h later cell-associated infectivity (only VSV) was determined.

Table 5. Effect of infection and double infection by fowl plaque and VSV on the uptake
and incorporation of labelled glucosamine and 2-deoxy-D-glucose

Infected with	^{3}H-Glucosamine Dpm in:		^{3}H-2-deoxy-D-glucose Dpm in:	
	Precipitate	Extract	Precipitate	Extract
Non-infected controls	1040	15300	2330	55300
Fowl plague	2800	33500	3450	98000
Fowl plague + VSV	1670	25000	2920	118000

Primary chick embryo cells in glucose-containing Earle's medium were infected either with fowl plague virus or VSV. Some fowl plague-infected cells were superinfected by VSV 2 h after primary infection. 3 h later a 45 min pulse with ^{3}H-glucosamine or ^{3}H-2-deoxy-D-glucose (5 µC/culture) was started. The radioactivity was determined in the acid insoluble material and in the acid extract.

Double infection by itself has no significant effect on VSV multiplication (ROTT et al., 1972).

Another possible explanation for the relative insensitivity of the multiplication of certain enveloped viruses against the treatment with the two sugar derivatives might be that these viruses tolerate wrongly glycosylated and/or wrongly processed glycoproteins within their envelope. It has been shown with VSV that it can easily incorporate envelope components of the paramyxovirus SV5 (CHOPPIN and COMPANS, 1970) or of murine leucemia and avian myeloblastosis virus (ZÁVADA, 1972), a phenomenon known as phenotypic mixing. Conditions have been worked out under which 2-deoxy-D-glucose does not prevent the formation of infectious VSV, although the viral glycoprotein synthesis is demonstrably disturbed. Such infectious VSV particles labelled by ^{3}H-amino acids can be shown to contain wrongly glycosylated and/or processed glycoproteins (SCHOLTISSEK et al., 1974b).

SAMSON and FOX (1974) described the effect of glucosamine on the synthesis of NDV glycoproteins in chick embryo cells. They found by polyacrylamide gel electrophoresis that the amino sugar selectively affects the synthesis of the virus-induced glycoproteins. Unfortunately these authors did not mention anything about the effects of glucosamine on the yield of infectious virus or on viral hemagglutinin in their system. Therefore it is unknown whether under their conditions infectious virus particles containing wrongly glycosylated viral envelope components are formed.

In summary, the disturbance of viral glycoprotein synthesis has 3 quantitatively clearly discernible effects on the multiplication of enveloped viruses:

1. Neither infectious nor physical particles are formed (fowl plague virus, Semliki Forest virus, avian sarcoma viruses).

2. Physical particles are formed which contain incorrect glycoproteins and which are not infectious (herpesviruses).

3. Physical particles are formed which contain incorrect glycoproteins and which in spite of this fact are infectious (VSV, possibly NDV).

From the data presented it can be concluded that the glycosylation plays an important role in the correct processing of the primary gene product.

IV. Effect of Glucosamine and 2-Deoxy-D-Glucose on Viral RNA Synthesis

A. Glucosamine

It has been mentioned in Chapter II A that the effect of glucosamine on the UTP pool of the host cell can be markedly enhanced if glucose in the culture medium is replaced either by fructose or pyruvate, which in contrast to glucose do not significantly compete with the uptake and metabolism of glucosamine. Thus in fructose-containing medium much lower doses of the amino sugar are needed for the inhibition of fowl plague and Semliki Forest virus replication. Under these conditions the multiplication of even NDV and VSV can be abolished by glucosamine (Scholtissek et al., 1975). A more thorough analysis of this effect on the multiplication of fowl plague and Semliki Forest virus has revealed that in fructose-containing medium the amino sugar interferes not only with the production of the glycoproteins, but also that synthesis of carbohydrate-free viral proteins is inhibited. The reason for this general suppression of synthesis of proteins is that under such conditions the production of viral RNA is abolished. With fowl plague virus the effect on the formation of virion RNA is more pronounced than the effect on complementary RNA synthesis. It might well be that the primary effect of glucosamine is on the production of virion RNA, which again is necessary as a template for the synthesis of complementary RNA. Since the effect on the synthesis of Semliki Forest virus RNA can be counteracted easily by adding uridine to the culture medium, but not after addition of other nucleosides, it is concluded that in fructose-containing medium glucosamine acts via a specific depletion of the UTP pool of the host cell. Under these conditions UTP becomes rate limiting for the production of viral RNA but not yet for cellular RNA. Correspondingly, glucosamine in fructose-containing medium has no significant effect on protein synthesis of non-infected cells. In spite

Table 6. Ratio of the specific radioactivity of CMP/UMP after labelling of fowl plague virus virion- and cellular RNA by ^{3}H-uridine

	Virion RNA	Cellular RNA
Dpm in CMP	9795	10665
Dpm in UMP	166615	292454
% of CMP in RNA	25.2	30.2
% of UMP in RNA	32.0	18.4
Spec. activity CMP/UMP	0.075	0.022

Chick embryo cultures were labelled with 50 μCi of ^{3}H-uridine, each, from 4 to 5 h after infection with fowl plague virus. Thereafter total RNA was extracted by phenol plus 0.5% sodium dodecyl sulfate at room temperature. After hybridization with a surplus of non-labelled complementary RNA and digestion with RNase the perchloric acid-precipitated (= virion RNA) as well as soluble material (=cellular plus about 10% complementary RNA) was digested by 0.5 M KOH at 65 °C for 45 min. Radioactivity was determined in CMP and UMP. For the calculation of specific radioactivity the base composition of fowl plague virus was taken from ROTT et al. (1965); that for total cellular RNA from SCHOLTISSEK (1965).

of this fact, cells infected either with fowl plague or Semliki Forest virus synthesize neither viral nor cellular proteins after treatment with the amino sugar. Thus the viral machinery designed to suppress cellular protein synthesis is still functioning in the presence of the amino sugar.

The specific effect of glucosamine in suppressing the production of viral RNA without influencing cellular RNA synthesis can be explained in at least two different ways: (1) either the cellular RNA polymerases (DNA-dependent) have a higher affinity to the common substrate UTP compared with viral RNA polymerases (RNA-dependent); or (2) two more or less independent UTP pools exist in chick embryo cells, one for cellular RNA synthesis which is not or is less affected by glucosamine, and a second one used for viral RNA synthesis, which is depleted easily by the amino sugar.

One explanation does not exclude the other. Some evidence in favor of the latter hypothesis has been derived from experiments with fowl plague virus-infected cells by comparing the labelling of cellular and viral RNA with ^{3}H-uridine. The ratio of the specific radioactivity in CMP to UMP after alkaline hydrolysis of the labelled cellular RNA is completely different from that of the virion RNA (Table 6). The radioactivity within the CMP-moiety is derived from the *de novo* synthesis of CTP via UTP. Thus the difference in the ratio of the specific radioactivity in CMP to UMP can be explained only if one assumes two different nucleoside triphosphate pools, one is used for cellular RNA synthesis, the other for viral RNA synthesis (DÄMMGEN and SCHOLTISSEK, in preparation).

The effect of glucosamine on the production of fowl plague or Semliki Forest virus envelope glycoproteins in contrast to its effect on viral RNA synthesis cannot be reversed by adding uridine to the culture medium (SCHOLTISSEK et al., 1975). This observation suggests that probably it is not the depletion of the UTP pool, but rather the accumulation of UDP-N-acetyl-glucosamine which inhibits the synthesis of the viral glycoproteins, because by the addition of uridine the UTP pool is refilled while more UDP-N-acetyl-glucosamine accumulates. Nothing is known yet about the mechanism by which UDP-N-acetylglucosamine could interfere with the synthesis of glyco-proteins.

B. 2-Deoxy-D-Glucose

As mentioned in Chapter III B, 2-deoxy-D-glucose does not inhibit the multiplication of NDV and VSV significantly in glucose-containing medium. The multiplication of both viruses is abolished, however, when glucose is replaced by pyruvate as an energy source and when even traces of glucose are excluded during the infection period, i.e. when virus grown in pyruvate medium or dialyzed against pyruvate medium is used for infection (Schol-tissek et al., 1974b). A more thorough analysis of this phenomenon has been performed with VSV. Labelling experiments with ^{3}H-amino acids have revealed that under these conditions none of the VSV-specific polypeptides are synthesized, although the inhibiting effect of virus infection on cellular protein synthesis is not abolished, i.e. in VSV-infected cells in the presence of 2-deoxy-D-glucose neither viral nor cellular proteins are produced. Viral RNA synthesis is highly reduced and the product (mainly complementary RNA) is rather unstable. From this observation it has been concluded that the primary effect of 2-deoxy-D-glucose under the conditions described is rather on viral RNA synthesis and/or stability, and as a consequence no viral proteins are produced. The observation that virus multiplication can be interrupted by the deoxy-sugar at any time after the onset of virus multiplication is in further agreement with this interpretation. If cells were infected with pyruvate-grown VSV and were grown in the presence of 2-deoxy-D-glucose, addition of glucose or mannose 4 h after infection caused an immediate rise in virus titer without any significant lag phase. Thus the early steps of virus multiplication, such as adsorption, penetration, and uncoating seem to occur normally in the presence of the antimetabolite. Furthermore, the genetic information is preserved under these conditions, since the effect of 2-deoxy-D-glucose is completely reversible. It is still obscure why the deoxy-sugar affects viral RNA synthesis and the yield of infectious virus only when glucose (or mannose) is completely omitted, even during the procedure of infection.

In summary, in fructose-containing culture medium, glucosamine depletes the host cell of UTP in such a way that the UTP pool becomes rate-limiting for viral RNA synthesis and no viral proteins are produced. RNA and protein syntheses of non-infected cells are, however, not affected under these conditions. Evidence has been presented that another pyrimidine nucleoside tri-phosphate pool is used for the synthesis of viral RNA which is separate from that used for the production of cellular RNA, and that glucosamine might influence these two pools to a different extent.

The production of such viruses as VSV which is influenced only very little by 2-deoxy-D-glucose in glucose-containing medium, can be inhibited by the deoxy-sugar when glucose or mannose is absent even during virus adsorption and penetration. Here the primary effect is again on viral RNA. Since 2-deoxy-D-glucose also has a striking effect on the labelling of UTP and RNA it should be checked whether under these conditions the cellular UTP pool also becomes rate limiting for viral RNA synthesis as shown in the case of glucosamine.

It has to be stressed here once more that although the sensitivity of virus multiplication against the two sugar derivatives is considerably increased in pyruvate- or fructose-containing medium, the toxicity of these two compounds is also increased (see Table 2). Thus, the reversibility of the inhibition has to be checked very carefully under each condition.

V. Other Effects of Glucosamine and 2-Deoxy-D-Glucose on Virus-Infected Cells

Various specific properties of different virus-infected cells can be affected by glucosamine and 2-deoxy-D-glucose and therefore seem to depend on the undisturbed synthesis of the viral glycoproteins. The properties to be discussed are the agglutinability of infected cells by Concanavalin A (BECHT et al., 1972), hemadsorption (MARCUS, 1962), and cell fusion from within (ROIZMAN, 1962; COHN, 1965; BRATT and GALLAHER, 1969). When chick embryo cells infected with an influenza A virus are treated with inhibiting doses of glucosamine in a glucose-containing medium, agglutinability by Concanavalin A does not occur (KALUZA et al., 1972). Under these conditions only the production of viral glycoproteins was disturbed. Under identical conditions cells infected by NDV have a normal yield of hemagglutinin (SCHOLTISSEK et al., 1975) and do not loose the agglutinability by the lectin (KALUZA et al., 1972). Hemadsorption of NDV-infected cells can be prevented by 2-deoxy-D-glucose (GALLAHER et al., 1973). Under the conditions employed, the yield of infectious NDV, however, was also suppressed by 90%. Since no data were presented on the yield of hemagglutinin in the presence of the sugar-antimetabolite, the results are difficult to interpret. There might already have been some suppression of hemagglutinin synthesis, and/or the incorporation of the (possibly wrongly glycosylated) viral glycoprotein into the (possibly altered) host cell membrane might be disturbed by 2-deoxy-D-glucose. Hemadsorption of influenza-infected cells also can be prevented by 2-deoxy-D-glucose (GANDHI et al., 1972).

Cell fusion caused by infection either with herpes simplex (GALLAHER et al., 1973) or pseudorabies virus (LUDWIG et al., 1974) can be prevented by 2-deoxy-D-glucose. There is also an inhibition of the fusion from within observed after infection with the paramyxoviruses NDV (GALLAHER et al., 1973; BORTFELDT, 1974) or SV5 (ROTT et al., 1974). According to BORTFELDT (1974), however, the fusion caused by infection with NDV is not completely inhibited but rather delayed by glucosamine or 2-deoxy-D-glucose. In contrast to the effect of the sugar antimetabolite against fusion from within, fusion from without is not prevented by 2-deoxy-D-glucose (GALLAHER et al., 1973).

In summary, all these observations point to the necessity of an undisturbed production of the viral and/or cellular glycoproteins in order to exhibit such effects as cell fusion and hemadsorption. The corresponding unglycosylated polypeptides seem to be unable to exert these effects.

VI. Concluding Remarks

Sugar derivatives, like glucosamine, N-fluoroacetyl-glucosamine, and 2-deoxy-D-glucose under controlled conditions specifically interfere with the production of enveloped viruses. Glucosamine acts via at least two different mechanisms. In glucose-containing medium it interferes only with the formation of the correctly glycosylated viral envelope components. The carbohydrate-free virus proteins and viral nucleic acids are produced normally. It has been suggested that this effect might be due to an accumulation of UDP-N-acetyl-glucosamine within the host cell.

If glucose is replaced by fructose, glucosamine depletes the UTP pool of the host cell in such a way that UTP becomes rate limiting for viral RNA synthesis leaving cellular RNA and protein synthesis still intact. Which of these effects are exhibited depends largely on the metabolic state of the host cell. This means that the amino sugar has to be taken up and metabolized optimally in order to become effective in either way. In this context, it would be interesting to study infected liver cells, since these cells take up and metabolize galactosamine most efficiently (KEPPLER et al., 1970) which has no effect in chick embryo cells (KALUZA et al., 1972). Thus in liver cells galactosamine might be a much better inhibitor than glucosamine.

In the system employing glucose-free medium which produces inhibition of viral RNA synthesis by depleting the host cells of UTP, the amino sugar should interfere not only with the multiplication of enveloped RNA viruses, but should also inhibit the production of other RNA-containing viruses. This hypothesis remains to be tested.

2-Deoxy-D-glucose is incorporated into viral glycoproteins. Since its effect is counteracted most efficiently by mannose it acts mainly as an antimetabolite to mannose. The most plausible explanation for its mechanism is that after its incorporation in place of mannose a further elongation and/or branching of the carbohydrate chain of the glycoproteins is not possible. In this way under- or unglycosylated viral envelope components with more or less faulty functions are produced.

Different viruses tolerate the production of incorrectly glycosylated envelope components to a different extent: in the case of influenza, toga- or avian sarcoma viruses, practically no physical or infectious particles are formed under inhibiting conditions. During herpesvirus maturation wrongly glycosylated envelope components are incorporated into virus particles which, however, lack infectivity. A rhabdo virus (VSV) tolerates the incorrect glycoproteins best, since they may be incorporated into viral particles which retain their infectivity.

Acknowledgement. I wish to thank my colleagues, Drs. R. ROTT, H.-D. KLENK, and R. T. FRIIS, for their valuable comments and criticisms of this paper.

References

BECHT, H., ROTT, R., KLENK, H.-D.: Effect of Concanavalin A on cells infected with enveloped RNA viruses. J. gen. Virol. **14**, 1–8 (1972)

BEKESI, J. G., BEKESI, E., WINZLER, R. J.: Inhibitory effect of D-glucosamine and other sugars on the biosynthesis of protein, ribonucleic acid, and deoxyribonucleic acid in normal and neoplastic tissues. J. biol. Chem. **244**, 3766–3772 (1969)

BEKESI, J. G., WINZLER, R. J.: The effect of D-glucosamine on the adenine and uridine nucleotides of sarcoma 180 ascites tumor cells. J. biol. Chem. **244**, 5663–5668 (1969)

BIELY, P., BAUER, S.: Metabolism of 2-deoxy-D-glucose by baker's yeast. I. Isolation and identification of phosphorylated esters of 2-deoxy-D-glucose. Coll. czech. chem. Commun. **32**, 1588–1594 (1967)

BIELY, P., BAUER, S.: The formation of guanosine diphosphate-2-deoxy-D-glucose in yeast. Biochim. biophys. Acta (Amst.) **156**, 432–434 (1968)

BORTFELDT, K.: Hemmung der durch die Paramyxoviren SV5 und NDV induzierten Fusion von BHK21-F-Zellen durch 2-Desoxy-D-Glucose und D-Glucosamin. Doctoral Thesis, Universität Gießen (1974)

BRATT, M. A., GALLAHER, W. R.: Preliminary analysis of the requirements of fusion from within and fusion from without by Newcastle disease virus. Proc. nat. Acad. Sci. (Wash.) **64**, 538–543

CHOPPIN, P. W., COMPANS, R. W.: Phenotypic mixing of envelope proteins of the parainfluenza virus SV 5 and vesicular stomatitis virus. J. Virol. **5**, 609–616 (1970)

COURTNEY, R. J., STEINER, S. M., BENYESH-MELNICK, M.: Effects of 2-deoxy-D-glucose on herpes simplex virus replication. Virology **52**, 447–455 (1973)

FISCHER, W., WEIDEMANN, G.: Die Umsetzung von 2-Deoxy-D-galactose im Stoffwechsel. II. Identifizierung der phosphorylierten Zwischenprodukte. Hoppe-Seylers Z. physiol. Chem. **336**, 206–218 (1964)

FLEMING, P.: Semliki Forest virus-chick embryo cell interactions. J. gen. Virol. **19**, 353–367 (1973)

GALLAHER, W. R., LEVITAN, D. B., BLOUGH, H. A.: Effect of 2-deoxy-D-glucose on cell fusion induced by Newcastle disease and herpes simplex virus. Virology **55**, 193–201 (1973)

GANDHI, S. S., STANLEY, P., TAYLOR, J. M., WHITE, D. O.: Inhibition of influenza viral glycoprotein synthesis by sugars. Microbios 5, 41–50 (1972)

HUNTER, E., FRIIS, R. T., VOGT, P. K.: Inhibition of avian sarcoma virus replication by glucosamine. Virology **58**, 449–456 (1974)

IBSEN, K. H., COE, E. L., McKEE, R. W.: A comparison of the respiratory inhibitions induced by D-glucose and 2-deoxy-D-glucose in Ehrlich ascites carcinoma cells. Cancer Res. **22**, 182–186 (1962)

KALUZA, G., SCHMIDT, M. F. G., SCHOLTISSEK, C.: Effect of 2-deoxy-D-glucose on the multiplication of Semliki Forest virus and the reversal of the block by mannose. Virology **54**, 179–189 (1973)

KALUZA, G., SCHOLTISSEK, C., ROTT, R.: Inhibition of the multiplication of enveloped RNA viruses by glucosamine and 2-deoxy-D-glucose. J. gen. Virol. **14**, 251–259 (1972)

KEPPLER, D. O. R., RUDIGIER, J. F. M., BISCHOFF, E., DECKER, K. F. A.: The trapping of uridine phosphates by D-galactosamine, D-glucosamine, and 2-deoxy-D-galactose. Europ. J. Biochem. **17**, 246–253 (1970)

KILBOURNE, E. D.: Inhibition of influenza virus multiplication with a glucose antimetabolite (2-deoxy-D-glucose). Nature (Lond.) **183**, 271–272 (1959)

KIPNIS, D. M., CORI, C. F.: Studies of tissue permeability. V. The penetration and phosphorylation of 2-deoxyglucose in the rat diaphragm. J. biol. Chem. **234**, 171–177 (1959)

KLENK, H.-D., SCHOLTISSEK, C., ROTT, R.: Inhibition of glycoprotein biosynthesis of influenza virus by D-glucosamine and 2-deoxy-D-glucose. Virology **49**, 723–734 (1972)

KLENK, H.-D., WÖLLERT, W., ROTT, R., SCHOLTISSEK, C.: Association of influenza virus proteins with cytoplasmic fractions. Virology **57**, 28–41 (1974)

Kohn, A.: Polykaryocytosis induced by Newcastle disease virus in monolayers of animal cells. Virology **26**, 228–245 (1965)

Laemmli, U. K.: Cleavage of structural proteins during the assembly of the head of the bacteriophage T4. Nature (Lond.) **227**, 680–685 (1970)

Lazdins, I., Haslam, E. A., White, D. O.: The polypeptides of influenza virus. VI. Composition of the neuraminidase. Virology **49**, 758–765 (1972)

Letnansky, K.: The influence of 2-deoxy-D-glucose on the nucleotide content of Ehrlich ascites carcinoma cells. Biochim. biophys. Acta (Amst.) **87**, 1–8 (1964)

Lewandowski, L. J., Bolognesi, D. P., Smith, R. E., Halpern, M. S.: Viral protein synthesis under conditions of glucosamine block in cells transformed by avian sarcoma viruses (1975) in press

Ludwig, H., Becht, H., Rott, R.: Inhibition of herpes virus-induced cell fusion by Concanavalin A, antisera and 2-deoxy-D-glucose. J. Virol. **14**, 307–314 (1974)

Marcus, P.: Dynamics of surface modification in myxovirus-infected cells. Cold Spr. Harb. Symp. quant. Biol. **27**, 351–365 (1962)

Morser, M. J., Burke, D. C.: Cleavage of virus-specified polypeptides in cells infected with Semliki Forest virus. J. gen. Virol. **22**, 395–409 (1974)

Plagemann, P. G. W., Erbe, J.: Transport and metabolism of glucosamine by cultured Novikoff rat hepatoma cells and effect on nucleotide pools. Cancer Res. **33**, 482–492 (1973)

Renner, E. D., Plagemann, P. G. W., Bernlohr, R. W.: Permeation of glucose by simple and facilitated diffusion by Novikoff rat hepatoma cells in suspension culture and its relationship to glucose metabolism. J. biol. Chem. **247**, 5765–5776 (1972)

Roizman, B.: Polykaryocytosis. Cold Spr. Harb. Symp. quant. Biol. **27**, 327–342 (1962)

Rott, R., Becht, H., Hammer, G., Klenk, H.-D., Scholtissek, C.: Changes of the surface of the host cell after infection with enveloped viruses. In: Negative strand viruses (eds. R. D. Barry and B. W. J. Mahy). London-New York: Academic Press 1974 (in press)

Rott, R., Becht, H., Klenk, H.-D., Scholtissek, C.: Interactions of Concanavalin A with the membrane of influenza virus infected cells and with envelope components of the virus particle. Z. Naturforsch. **27**b, 227–233 (1972)

Rott, R., Saber, S., Scholtissek, C.: Effect on myxoviruses of mitomycin C, actinomycin D, and pretreatment of the host cell with ultra-violet light. Nature (Lond.) **205**, 1187–1190 (1965)

Rott, R., Scholtissek, C., Klenk, H.-D., Kaluza, G.: Intrinsic interference between different enveloped RNA viruses. J. gen. Virol. **17**, 255–264 (1972)

Samson, A. C. R., Fox, C. F.: Selective inhibition of Newcastle disease virus-induced glycoprotein synthesis by D-glucosamine hydrochloride. J. Virol. **13**, 775–779 (1974)

Schmidt, M. F. G., Schwarz, R. T., Scholtissek, C.: Nucleoside diphosphate derivatives of 2-deoxy-D-glucose in animal cells. Europ. J. Biochem. **49**, 237–247 (1974)

Scholtissek, C.: Synthesis of an abnormal ribonucleic acid in the presence of proflavine. Biochim. biophys. Acta (Amst.) **103**, 146–159 (1965)

Scholtissek, C.: Detection of an unstable RNA in chick fibroblasts after reduction of the UTP pool by glucosamine. Europ. J. Biochem. **24**, 358–365 (1971)

Scholtissek, C.: Influence of glucosamine on the uptake of nucleosides by chick fibroblasts and on the incorporation into RNA. Biochim. biophys. Acta (Amst.) **277**, 459–465 (1972)

Scholtissek, C., Kaluza, G., Rott, R.: Stability and precursor relationships of virus RNA. J. gen. Virol. **17**, 213–219 (1972)

Scholtissek, C., Kaluza, G., Schmidt, M., Rott, R.: Influence of sugar derivatives on glycoprotein synthesis of enveloped viruses. In: Negative strand viruses (eds. R. D. Barry and B. W. J. Mahy). London-New York: Academic Press 1974a (in press)

Scholtissek, C., Rott, R., Hau, G., Kaluza, G.: Inhibition of the multiplication of vesicular stomatitis and Newcastle disease virus by 2-deoxy-D-glucose. J. Virol. **13**, 1186–1193 (1974b)

Scholtissek, C., Rott, R., Klenk, H.-D.: Two different mechanisms of the inhibition of the multiplication of enveloped viruses by glucosamine. Virology, in press (1975)

Schwarz, R. T., Klenk, H.-D.: Inhibition of glycosylation of the influenza virus haemagglutinin. J. Virol. **14**, 1023–1034 (1974)

STEINER, S., COURTNEY, R. J., MELNICK, J. L.: Incorporation of 2-deoxy-D-glucose into glycoproteins of normal and simian virus 40-transformed hamster cells. Cancer Res. **33**, 2402–2407 (1973)

STEINER, S., STEINER, M. R.: Incorporation of 2-deoxy-D-glucose into glycolipids of normal and SV 40-transformed hamster cells. Biochim. biophys. Acta (Amst.) **296**, 403–410 (1973)

WOODWARD, G. E., HUDSON, M. T.: The effect of 2-deoxy-D-glucose on glycolysis and respiration of tumor and normal tissues. Cancer Res. **14**, 599–605 (1954)

ZÁVADA, J.: Pseudotypes of vesicular stomatitis virus with the coat of murine leucaemia and of avian myeloblastosis viruses. J. gen. Virol. **15**, 183–191 (1972)

Subject Index

Italized numbers refer to pages where the term appears in a table or a figure.

Springer-Verlag
Berlin
Heidelberg
New York

Current Topics in Microbiology and Immunology

Vol. 66
With 24 figures
III, 125 pages. 1974
Cloth DM 48,–; US $20.70

Contents
S. Ferrone, M.A. Pellegrino,
M.P. Dierich, R.A. Reisfeld:
Expression of Histocompati-
bility Antigens during the
Growth Cycle of Cultured
Lymphoid Cells
D.G. Braun, J.-C. Jaton: Homo-
geneous Antibodies: Induction
and Value as Probe for the
Antibody Problem.
E. Wintersberger: Nucleic Acid
Synthesis in Yeast.

Vol. 67
With 7 figures
III, 161 pages. 1974
Cloth DM 68,–; US $29.30

Contents
H. Wigzell, P. Häyry:
Specific Fractionation of Immu-
nocompetent Cells. Applications
in the Analysis of Effector Cells
Involved in Cell Mediated Lysis.
J.L. Reissig: Decoding of
Regulator Signals at the Microbial
Surface. A.J. Cunningham:
Predicting what Antibodies an
Antigen will Induce; The
Inadequacy of the Determinant
Model. J.E. Larsh,
N.F. Weatherly: Cell Mediated
Immunity in Certain Parasitic
Infections.

Vol. 68
With 23 figures
III, 174 pages. 1974
Cloth DM 74,–; US $31.90

Contents
P.S. Sarma, A.F. Gazdar:
Recent Progress in Studies of
Mouse Type-C Viruses.
H.-D. Klenk: Viral Envelopes
and their Relationship to
Cellular Membranes.
R. Goldstein, J. Lengyel, G. Pruss,
K. Barrett, R. Calendar, E. Six:
Head Size Determination and the
Morphogenesis of Satellite Phage
P4. B.R. McAuslan,
R.W. Armentrout: The Bio-
chemistry of Icosahedral Cyto-
plasmic Deoxyviruses.
R.M. Frankling: Structure and
Synthesis of Bacteriophage PM2
with Particular Emphasis on
the Viral Lipid Bilayer.

Vol. 69
With 12 figures
III, 200 pages. 1975
Cloth DM 78,–; US $33.60

Contents
A.C. Frazer, R. Curtiss:
III, Production. Properties, and
Utility of Bacterial Minicells. –
C.R. Pringle: Conditional Lethal
Mutants of Vesicular Stomatitis
Virus. – I. Horak: Translation
of Viral RNA in Cell – Free
Systems from Eukaryotes. –
R.A. Finkelstein: Immunology of
Cholera.

Current Topics in Microbiology and Immunology

Reprint from

Vol. 70

Theoretical Aspects of Structure and Assembly of Viral Envelopes

H. A. Blough and J. M. Tiffany

Springer-Verlag   Berlin · Heidelberg · New York   1975

Current Topics in Microbiology and Immunology

Reprint from Vol. 70

Latent Herpes Simplex Virus and the Nervous System

J. G. Stevens

Springer-Verlag   Berlin · Heidelberg · New York   1975

Current Topics in Microbiology and Immunology

Reprint from

Vol. 70

Temperature-Sensitive Mutants
of Herpesviruses

P. A. Schaffer

Springer-Verlag Berlin · Heidelberg · New York 1975

Current Topics in Microbiology and Immunology

Reprint from Vol. 70

Inhibition of the Multiplication of Enveloped Viruses by Glucose Derivatives

C. Scholtissek

Springer-Verlag   Berlin · Heidelberg · New York   1975

GPSR Compliance
The European Union's (EU) General Product Safety Regulation (GPSR) is a set
of rules that requires consumer products to be safe and our obligations to
ensure this.

If you have any concerns about our products, you can contact us on

ProductSafety@springernature.com

In case Publisher is established outside the EU, the EU authorized
representative is:

Springer Nature Customer Service Center GmbH
Europaplatz 3
69115 Heidelberg, Germany